# THAT BUCK ROGERS STUFF

# THAT BUCK ROGERS STUFF

JERRY POURNELLE

MAN REACHES
STARS

That Buck Rogers Stuff
Jerry Pournelle

Published by Chaos Manor™



Cover art by Raul Garcia Capella.



ISBN: 978-952-7303-25-2

## PRAISE FOR *THAT BUCK ROGERS STUFF*

In the early 70s the American zeitgeist was turning to crap—kinda like today. The Club of Rome's *The Limits to Growth* called for a halt to technological advance and a reining in of industrial production, Paul Ehrlich's *The Population Bomb* predicted mass starvation by the late 70s, while Stuart Brand was championing "Small is Good" and the end of Western Civilization as we know it.

The only ray of hope was beaming from an up-and-coming Science-Fiction author, Jerry Pournelle, whose *Galaxy Magazine* column "A Step Farther Out" proclaimed loudly that the only limit to the future was Man's own nerve. Progress and technology were mankind's greatest hope, and if we invested wisely we could all survive with style. Space was the next frontier, and we could conquer that and our fears if only we trusted in ourselves.

—John F. Carr

I first read *That Buck Rogers Stuff* in about 1980. I was blown away by Jerry's casual mastery of a dozen disciplines, as well as the folksy way he could convey insanely complex concepts. He was simply one of the greatest essayists in the field. I hope you've never read this collection, because if you haven't, you have hours of serious pleasure ahead. If you have... I bet you've forgotten just how good he was. Remind yourself. You won't be disappointed.

—Steven Barnes

Jerry Pournelle was one of the sharpest and most imaginative writers we've ever had, and these essays are like having long conversations with him. Witty, surprising,informed and convincing, this book is one you'll re-read many times, always getting more out of it.

—Tim Powers

# Contents

# ACKNOWLEDGMENTS

This book would not exist but for the investment of time and effort on the part of many people, to whom I wish to pay credit:

Larry Niven graciously provided both the introduction to this volume and permission to include the collaboration on the background of *The Mote in God's Eye.* The cover artwork was ably drawn by Ray Capella, who had very little time and only a hint of what I desired to work with.

Marty Massoglia and Craig Miller put me in contact with the artist and printer, respectively; their enthusiastic support for the project helped turn it into reality. Likewise Nick Smith, who also typed a portion of the text. John McLane produced the interior diagrams expertly and rapidly. Some crucial technical assistance was provided by James D. Jones of The Type Shoppe, Glendale.

My selection of articles for the book was greatly influenced by the ideas of Andrew J. Galambos, as disclosed through his lectures at the Free Enterprise Institute. However, this should not be considered an endorsement by him of the articles herein, nor of the concept of "Buck Rogers." The value I have received from certain concepts is reflected here, so proper crediting is due.

The inspiration for the book came partially from William Cox, who expressed a desire to see wider circulation of "Survival With Style" and similar articles. I wish also to thank the past few Boskone committees, who also inspired the book, having done some themselves. It was reassuring to know that it *really* was possible.

I am grateful to Dr. Pournelle for his help in preparing this collection, and especially for having written the articles in the first place. Had there been space enough, I would have eagerly included a dozen more.

—Gavin Claypool

# AN INTRODUCTION TO JERRY POURNELLE

by Larry Niven

The most successful collaboration team in science fiction history is Jerry Pournelle and me, from my viewpoint. (You get to make your own decision.) It was not an accident. Jerry came looking for me.

He had already lived an unusually busy lifetime. Jerry had fought in Korea as a Lieutenant. He was in the space industry when there wasn't any, when parts and working time for the rockets had to be stolen from mundane projects. He was carving out the discipline of Space Medicine the day a prospective astronaut froze in panic, in a spacesuit, in a chamber at 1500°F. Jerry had to go in after him, soaking wet and armed with wet blankets and a secret I pass on, just in case it ever happens to you: "Don't breathe."

He taught college for awhile in central L.A., driving in rush hours, an hour to work and an hour and a quarter home, five days a week, with a dictaphone in his car to keep him sane. (I find the concept horrifying.) When that ended he faced a thorny decision: a high-paying job offer that would have moved him and his wife, Roberta, and four kids, to another city, versus an urge to write science fiction.

That was no easy choice. Jerry's been a fan much longer than I have. He grew up a starry-eyed daydreamer (like me) who did his homework (whereas I didn't). (I do now. He makes me.)

He and Roberta decided to give him the chance. He tried three espionage novels, under the name "Wade Curtis," while he looked about for a collaborator who already knew how to write science fiction. He may have considered an old friend, Robert Heinlein, for up to ten minutes. But with some men you *know* that their dream-worlds are too personal to intrude upon. (I've been asked often enough: how can you *possibly* collaborate? Not just with Jerry, though that happens too; how can you share a world-concept? How can you compromise your personal daydreams? Well, it's possible, but not for everyone.)

Another old friend, Poul Anderson, had already written collaborations. But Poul was living in San Francisco. A brilliant newcomer, a nuts-and-bolts SF writer who had already written a collaboration novel with David Gerrold, lived less than half an hour away. It was certainly worth talking about.

"I'll make you rich and famous," he told me.

"I'm already rich," I pointed out.

"Okay. You make me rich. I'll make you famous."

"Two things about a collaboration," I said. "First, one of us has to have absolute veto power in case of an unresolved disagreement. Second, one of us has to rewrite the whole damn thing—put all of it through a typewriter again, once we've got a near-final draft. It's the only way I can think of to smooth out the jarring inconsistencies in style." I'd learned *something* from *The Flying Sorcerers.*

He said, "You get the veto. I'll do the full rewrite." And I knew we were going to write a novel. So we sat down with a lot of coffee and brandy (I taught him that sin) and talked, and made notes, and talked.

He couldn't accept the Known Space universe. The treatment of history was too unrealistic, he said. Oh, all right, we'd work within his own future history. He spent some time showing it to me. It wasn't as exciting as Known Space… in fact, it had a thousand worlds with nothing but humans in it, just like the Foundation series. I took another look at the Alderson Drive, and talked with Dan Alderson. It looked like I could drop a whole worldful of intelligent aliens right into the middle of Jerry's Second Empire, without anyone having suspected their existence for a thousand years. Wouldn't *that* be fun! And I had a carefully designed alien left over from a novella that bogged down two-thirds of the way through….

You really should have heard him (and some of you did) when it came time to do the final rewrite for *The Mote in God's Eye.* Oh, he was going to do it! He never considered backing out. But it was four times as long as what we'd thought we were writing.

Did my conscience bother me? Certainly not. Think of it as an apprenticeship fee. Besides, I rewrote *Inferno*. Which was a quarter the length of *Mote*.

The Rich and Famous part is working out fine. *Lucifer's Hammer* went at paperback auction for $236,500, and it's getting enormous publicity; the prize was a two-page spread in the New York Times, at $10,000—which is seven times the advance on my first book, and five times the advance on *Ringworld*. But what works best is a coincidence, something Jerry couldn't have planned on.

There were gaps in our skills… and each gap in one of us matched an area of expertise in the other. I have to use the past tense here, because we've each learned a lot from the other. We did not have to push *Lucifer's Hammer* through the typewriter again. (And the book was our editor's idea. He would have gotten the job.) We've learned to match our styles. I can handle some politics; he can handle some aliens. He can round out a character he hates; I can indulge in political discussions. The character holding hard to his sanity and his purpose in the face of adversity is Jerry's; but not always. The half-mad character is mine; but not always.

Jerry's own career? You're holding part of it. For some years he's been writing science articles not only for *Galaxy*, but also for *Twin Circles*, a Catholic magazine with heavy circulation. We are both fanatics (though I am a lazy fanatic) on certain subjects, and this is one: that science solves more problems than it creates; that without expanding knowledge, we are lost. The mundanes who read *Twin Circles* need that knowledge.

He wrote two Laser Books. He wrote *Birth of Fire* in a week; finished on a Saturday and staggered off to a poker game. He lost a fair amount, but he still figured he was making more per hour that afternoon. *West of Honor* took him about the same time. When Laser was folding, our editor, Bob Gleason, attended a distributors' convention. "These are evil, foul-mouthed people who don't read books," he says. "I overheard someone at the next table saying, 'Those ****sucking Laser Books weren't ever worth dog****. The only ones that made anything back were written by somebody... Pool something....' I turned around and asked. Poul? Pohl? Turned out to be Pournelle"

Jerry writes in two future histories: the one that ends with *Mote* (so far) and another that deals with higher technology and a slower-than-light universe. *The Mercenary* is selling extremely well and has been throwing off side-effects: board games, Masquerade costumes, fan clubs. He spent enough man-hours on a novel, *The Last King of Atlantis*, to have written *Mote* again, and he's finally figured out why he can't write it. Unfortunately he's right. Perhaps it will restore his characteristic humility. The way *Lucifer's Hammer* has been selling, he needs that.

# THE BEST TRACK RECORD

In my columns and lectures I often tell of the marvels about to be poured forth from technology's cornucopia. I describe a world of the future with colonies in space, minerals brought from the asteriods, a world-wide standard of living at least as high as what we in the United States enjoy now; and I am careful to say that I am not describing dreams. This is the world as it can be made, as we already know how to make it. We can do it, I say. And it doesn't even cost much: a few more cents out of each tax dollar.

I gave that lecture in Salt Lake City recently. (Salt Lake City is the only place I've ever visited where the words "wild life" refer exclusively to ecological phenomena.) After the question period I stood talking with some of my audience, and a young lady asked a very serious question. "You tell us about all the benefits technology can bring us, and you say you only need a little more money to accomplish all these marvels. I'm not an engineer. I don't even understand about half of what you said in there. I'd like to believe you, but—how do I know you can do it if we give you the money?"

It was asked in all sincerity. Most of my audience wasn't technically trained, and indeed in this case weren't even very familiar with science fiction, and though I've evidence they were entertained, I knew too that some of the things I'd talked about were unfamiliar. I could probably convince an engineer or mathematician or economist that my forecasts make sense; but how to prove to a bright young English teacher that I wasn't just blowing smoke, that it wasn't all just promises, promises?

I had to say something, and I heard myself saying this: "Of those who make you promises, which group has a better track record for keeping them: technologists or politicians?" She seemed satisfied; and later I reflected on just what I'd said. It makes more sense that I knew.

Remember twenty years ago when the politicians and "social scientists" were saying that if only they had as much money as the Defense Department, they would transform America into Paradise? Well, they've got what they asked for and a lot more. I can recall when Congress dared not bring in a budget larger than the "barrier" figure of $100 billion. Now there's no obvious stopping point short of a trillion—and not much of the increase went to Defense. Do we live in Paradise?

In 1958 some of us said that if we could have about 3% of the national budget we could put men on the Moon and go to the planets. We said that the nation would reap great benefits from communications and weather satellites, and that the ferment of high-technology enterprises generated by the space program would have unforeseeable effects of enormous benefit to all. Have those promises been kept?

In the 50's advocates of "federal aid to education" were saying that if we merely shoveled a bit more money into education we'd not only see that every Johnny could read, but produce a generation fit to live in "the atomic age." Well, education certainly can't complain that it didn't get far more than was asked for (asked for then; not as much as is wanted now, of course); but would anyone like seriously to argue that we have fewer problems with the schools now than we did then?

And no: I do not mean this as a condemnation of educators and politicians and social scientists. I do not mean to imply that there may not be serious problems not foreseen by those forecasters of the 50's. I don't even mean to condemn those who tell us what's needed is still more money for education and social services and the like. I do mean this: of those who have said they could produce certain results given certain investments, who has the best track record? And yes, I know about cost over-runs (back in my aerospace days I was mildly famous for Pournelle's Law of Costs and Schedules, namely, "Everything takes longer and costs more," a dictum discovered independently by myself and Poul Anderson).

I could even tell you horror stories of my own. Some of them aren't the engineers' fault, though. Freeman Dyson told me about the laser target, the one placed on the Moon by Neil Armstrong: essentially a box full of glass cubes. They asked the instrument makers at Princeton what it would cost, and were told a couple of thousand dollars at most; by the time the competitive bid process was done, the cost was about a quarter of a million. But sometimes the engineers and technology managers seriously underestimate their costs, and seriously overestimate the results. Sometimes they build outright failures, bridges that fall down and airplanes that don't fly very well. But be honest. How often have we been given an order of magnitude more money and failed to produce the promised result? Or any result at all? And how many political programs do you know of that cost ten times as much as estimated, are seemingly eternal in duration, and produce no measurable result at all?

Choose your own examples; I'd not like to pick on your favorite project for social improvement. I do recall Dr. Samuel Johnson on the subject.

Boswell: "Then, sir, you laugh at schemes for social improvement?"

Johnson: "Why, sir, most schemes for social improvement are very laughable things."

And yet in my youth no one laughed when we were told that for $200 billion—not annually, but just $200 billion—we could transform the world, and they did indeed laugh when told that we could go to the Moon at any price whatever.

I rest my case.

# HALFWAY TO ANYWHERE

One of my rivals in the science-writing field usually begins his columns with a personal anecdote. Although I avoid slavish imitation, success is always worth copying. Anyway, the idea behind this column came from Robert Heinlein, and he ought to get credit for it.

Mr. Heinlein and I were discussing the perils of template stories: interconnected stories that together present a future history. As readers may have suspected, many future histories begin with stories that weren't necessarily intended to fit together when they were written. Robert Heinlein's box came with "The Man Who Sold the Moon." He wanted the first flight to the Moon to use a direct Earth-to-Moon craft, not one assembled in orbit—but the story had to follow "Blowups Happen" in the future history.

Unfortunately, in "Blowups Happen" a capability for orbiting large payloads had been developed. "Aha," I said. "I see your problem. If you can get a ship into orbit, you're halfway to the Moon."

"No," Bob said. "If you can get your ship into orbit, you're halfway to *anywhere*."

He was very nearly right.

Space travel isn't a matter of distances, it's a question of velocities. Now most space systems designs begin with rough-cut estimates of present and near-term predicted technological capabilities; and one of the best measures used in design analysis is called *delta-v*. This is engineer talk for a change in velocity, and comes from the general mathematical symbol for change, the Greek letter $\Delta$ (delta). Delta-v, written $\Delta v$, is the total velocity change a ship can make.

The nice part about delta-v is that for rough analysis it doesn't matter how you expand your fuel. You can burn it all up at once, or make a whole series of velocity changes: the sum of delta-v achieved will be the same. Moreover, the total delta-v can be calculated from the *specific impulse* (a measure of efficiency) of the fuel used and the fraction of the total ship weight that's made up of fuel. No other numbers are needed, not even total ship's weight. Given the total delta-v, you can determine what kind of missions the ship can perform.

The other nice feature is that delta-v requirements for any journey in the solar system can be calculated from well-known parameters: mass of the Sun, masses of the planets you're leaving and going to, and the distances of the planets from the Sun. There are a lot of possible refinements, but rough estimates of delta-v requirements for any minimum-energy journey can be run off on a pocket computer in no time.

The least-costly method of long-distance space travel involves transfer orbits, sometimes called *Hohmann orbits* after the German architect Dr. Walter Hohmann who first calculated the energy requirements to get from place to place in the solar system. Hohmann's book, *The Attainability of the Celestial Bodies*, was published in the mid-30's and was a very important book indeed, because it showed that space travel really was possible with chemical rocket fuels.

Unfortunately, as Willy Ley noted in *Rockets and Space Travel*, Hohmann's book is nearly unreadable, combining Germanic scholarly thoroughness, unfamiliar subject matter, lots of mathematics, and an unnervingly complex style. Despite that, his work remains important and the transfer orbits he described are the only feasi-

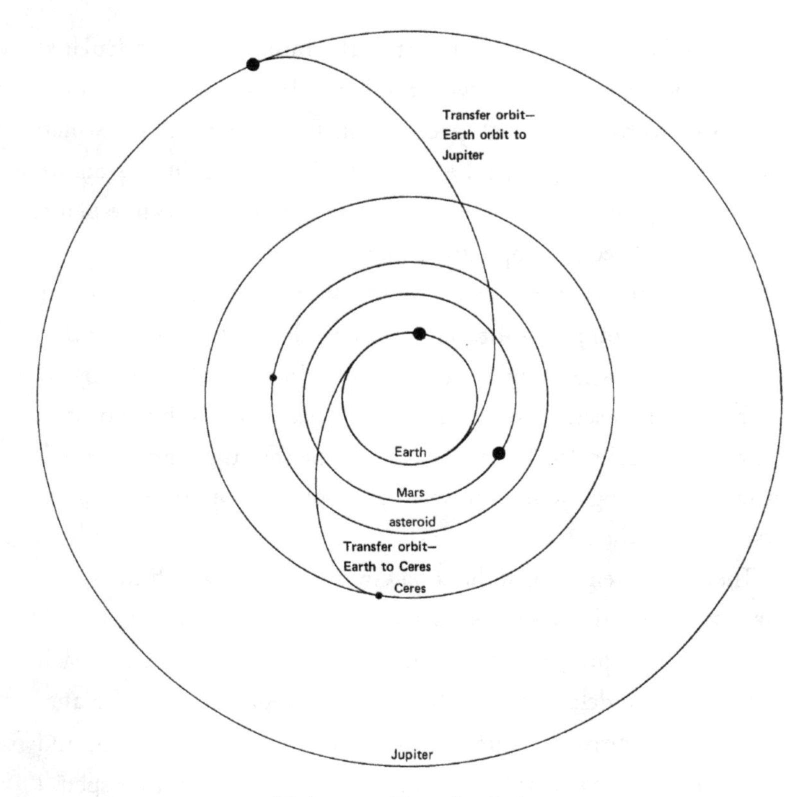


*Hohmann Transfer Orbits*

ble methods of getting to other planets from Earth with chemical rockets.

In Hohmann orbits, the starting planet at launch time and the target planet at time of journey's end must be precisely opposite each other with the Sun between (see figure 1). Naturally, then, the trip begins when the target planet hasn't yet reached opposition—these journeys can start only at certain times. The ship departs on a trajectory that carries it into a highly elliptical orbit with one end of the ellipse just touching the orbit of the origin planet and the other touching the orbit of the target planet.

The delta-v required for Hohmann trips to various places is shown in table 1. In every case it is assumed that the starting point is not on Earth, but in orbit around Earth. The numbers were calculated for me by Dan Alderson, who programs JPL's computers and is usually concerned with real spacecraft such as Pioneer and Mariner; they're quite accurate given the model used. For those interested, we assume the planets have circular orbits and all lie in the same plane, and use conic-section approximations.

The first important number is the fly-by delta-v requirement. This assumes you just want to get close to the target, and after that you don't care what happens to the ship. In the real world, fly-by probes can be useful afterwards: the Pioneer series Jupiter probes, for example, rounded Jupiter in such a way that they used Jupiter's attraction to fling them on toward' other planets, or out of the solar system altogether.

There was even a possibility of a Grand Tour, in which the spacecraft approached Jupiter, Saturn, and then either flew past both Uranus and Neptune, or went directly from Saturn to Pluto, each time using the delta-v gained from a close approach to one planet to get to the next. Congress wouldn't fund the Grand Tour, and that opportunity is lost for our lifetimes because it takes a special configuration of the outer planets.

The Pioneer probes carry the famous gold plaque with a code showing the origin of the spacecraft and line drawings of human beings, male and female, on the assumption that someday they may be picked up by beings in another star system. Since the probe will leave the solar system with a velocity of only a few kilometers per second, and must cross trillions of kilometers before there's any possibility of it being found, we don't have to worry much about the aliens using it to track us back to Earth and conquer us. By that time—if interstellar travel is possible—we'll have it.

It happens that I was present when that plaque—called "The Praque" by the TRW technicians who build Pioneer—came about. NASA held a big press briefing at TRW, a dog and pony show for science reporters. The NASA, JPL, and TRW scientists concerned with Pioneer described the experiments aboard, and one happened to mention that Pioneer would definitely leave the solar system forever.

One of the reporters present was Eric Burgess, who with Arthur Clarke founded the British Interplanetary Society back in the 40's. Eric became very thoughtful, and later that afternoon spoke to Carl Sagan of Cornell and some of the others in charge of Pioneer, pointing out what a unique opportunity this was to send a message to anyone "out there." It might take a long time to arrive, but at least it was going. The idea caught on, and within a week the plaque was designed and installed.

Then, of course, came the complaints about the "dirty pictures" of nude men and women, but that's another story.

Table 1 shows in addition to fly-by delta-v requirements, the delta-v you'd need to get into some kind of orbit around the planet: the bare minimum for capture, and a circular orbit from which you could land or observe closely. You can see the numbers come out at reasonable values, except when you're trying to get very close to the Sun. One important number is the Sun's escape velocity. If you have that much delta-v capability, you can get to other stars—anywhere, for practical purposes. It is important to note, though, that the table assumes you don't start from Earth, but from *orbit around Earth.*

Since you need 7.6 km/sec delta-v to get into Earth orbit in the first place, Bob Heinlein's top of the head remark was very close to correct. Earth orbit is halfway to anywhere.[1]

In other words, the first step is the hard one. If you can get into Earth orbit, you can get most anywhere you want to go. Unfortunately the disintegrating totem poles we now use to get into orbit

[1] Quibblers will know that you'd have to stay in the plane of the ecliptic or use a lot more energy to get out of it—and that the galaxy itself has a very high escape velocity, on the order of 100 km/sec from here.

Table 1: Delta-v's to Celestial Bodies

| Target | Average Distance from Sun (kilometers) | Fly-by Delta-v (km/sec) | Marginal Capture Delta-v (km/sec) | Circular Capture Delta-v (km/sec) |
|---|---|---|---|---|
| Sun | | 21.249 | — | 200.786 |
| Mercury | 57,900,000 | 5.580 | 11.874 | 13.104 |
| Venus | 108,000,000 | 3.555 | 3.905 | 5.470 |
| Earth | 148,000,000 | 3.280 | 3.280 | 3.280 |
| Mars | 228,000,000 | 3.661 | 4.320 | 5.535 |
| Asteroid | 300,000,000 | 4.378 | 8.320 | 8.320 |
| Ceres | 414,000,000 | 4.691 | 9.530 | 9.530 |
| Jupiter | 778,000,000 | 6.322 | 6.583 | 10.315 |
| Saturn | 1,430,000,000 | 7.293 | 7.691 | 11.143 |
| Uranus | 2,870,000,000 | 7.981 | 8.469 | 11.277 |
| Neptune | 4,500,000,000 | 8.248 | 8.575 | 11.116 |
| Pluto | 5,910,000,000 | 8.363 | 8.841 | 10.972 |
| Escape | infinite | 8.748 | — | — |

Values for Sun are very close approach and circular orbit at surface. Value for Earth is marginal delta-v needed to escape Earth's gravitational effect. Asteroid capture values are large because the asteroids have essentially no mass, and thus do not aid appreciably in an attempt to catch up with them after arriving at their orbital distance.

are just too cumbersome and expensive to make space travel routine. Worse, they use up nearly all their total delta-v getting into orbit—and the rocket is thrown away, hundreds of millions of bucks into the drink.

The upcoming Shuttle will help and is sorely needed, but there's a system even better than that. The concept I'm about to describe can use old rocket boosters over and over again—in fact, the rocket motor never leaves the ground. Only payload goes up.

This magic feat is performed by lasers. The basic design of the system comes from A. N. Pirri and R. F. Weiss of Avco Everett research laboratories. What they propose is an enormous ground-based laser installation consuming about 3000 megawatts. In practice, there would probably be a number of smaller lasers feeding into mirrors, and the mirrors would then concentrate the beam onto one single launching mirror about a meter in diameter. This ground station boosts the spacecraft—the ships themselves carry no rocket motors, but instead have a chamber underneath into which the laser beam is directed.

The spacecraft weigh about a metric ton (1000 kilograms or 2200 pounds) and are accelerated at 30 g's for about 30 seconds—that puts them in orbit. While the capsule is in the atmosphere the laser is pulsed at about 250 hertz (cycles per second when I was in school). Each pulse causes the air in the receiving chamber to expand and be expelled rapidly. The chamber refills and another pulse hits—a laser-powered ramjet. For the final kick outside the atmosphere the laser power is absorbed directly in the chamber and part of the spacecraft itself is ablated off and blown aft to function as reaction mass. Of the 1000 kg starting weight, about 900 kg goes into orbit.

Some 80 metric tons can be put into orbit each hour at a total cost of around 3000 megawatt-hours. Figuring electricity at 3 cents a kilowatt-hour, that's $90,000—about $1.10 a kilogram—for fuel costs. Obviously there are operating costs and the spacecraft aren't free, but the whole system is an order of magnitude more economical than anything we have now.

Conventional power plants cost something like $300 a kilowatt; a 3000-megawatt power plant would run close to a billion dollars in construction costs. However, when it isn't being used for space launches it could feed power into the national grid, so some of that is recovered as salable power. The laser installation might easily run $5 billion, and another $5 billion in research may be needed.

The point is that for an investment on the order of what we put out to go to the Moon, we could buy the research and construct the equipment for a complete operating spaceflight system, and then begin to exploit the economic possibilities of cheap space travel.

Many benefits would accrue to an economical system of putting payload into orbit. Some are commercial—*e.g.*, the manufacture of materials that can only be made in gravity-free environments. Others are not precisely commercial, but highly beneficial. For example, the power/pollution problem is enormously helped. Solar cells can collect sunlight that would have fallen onto the Earth. They convert it to electricity and send it down from orbit by microwave. That's fed into the power grid, and when it's used it becomes heat that would have arrived here anyway—the planetary heat balance isn't affected.

Interestingly enough, it's now believed that orbiting solar power plants can be economically competitive with conventional plants, provided that we get the cost of a kilogram in orbit down to about $45. The laser launch system could power itself.

We don't even have to build a permanent power plant to get the laser launcher into operation. There are a lot of old rocket motors around, and they're very efficient at producing hot ionized gases. Hot ionized gas is the power source for electricity extracted by magnetohydrodynamics (MHD). MHD is outside the scope of this article, but basically a hot gas is fed down a tube wrapped with conducting coils, and electricity comes out. MHD systems are about as efficient as turbine systems for converting fuel to electricity, and they can burn hydrogen to reduce pollution.

The rocket engines wouldn't last forever, and it takes power to make the hydrogen they'd burn—but we don't have to use the system forever. It needn't last longer than it takes to get the big station built in space and start up a solar-screen power plant.

None of this is fantasy. The numbers work. Avco has done some experiments with small-scale laser-powered "rockets" and they fly.

There are no requirements for fundamental breakthroughs, only a lot of development engineering, to get a full-scale working system.

Laser launchers are at about the stage that rockets were at circa 1953. Fifteen years and less than $20 billion would do the job and we'd have a system to get nearly anything we wanted to have out there into orbit.

That doesn't seem like very much to get halfway to anywhere.

# HOW LONG TO DOOMSDAY?

"While you are reading these words four people will have died from starvation. Most of them children."

Thus opens Paul Ehrlich's *The Population Bomb*.

"It seems to me, then, that by 2000 A.D. or possibly earlier, man's social structure will have utterly collapsed, and that in the chaos that will result as many as three billion people will die. Nor is there likely to be a chance of recovery thereafter...."

Thus closes a popular article by Dr. Isaac Asimov, perhaps the best-known science writer in America.

It would not be hard to multiply examples of doom-crying among science-fiction writers. There are dozens of stories describing life in these United States in the year 2000 as poor, nasty, brutish, and short—although hardly solitary, as Hobbes would have had it.

Much of this doom-saying springs from three books: Ehrlich's work previously mentioned and two outputs from MIT: *World Dynamics* and *The Limits to Growth*. All are essentially mathematical trend projections, with the MIT studies employing detailed computer models.

Strangely, intellectuals including SF writers have a lot of confidence in these models, although they have very little in the ability of social or physical scientists to save us. It's almost impossible to overestimate the influence of these three books. Writers make predictions based on them; teachers quote fourth-hand sources which quote the original studies. They have become "conventional wisdom" for the young.

I don't have that much faith in any predictions; perhaps it's time to look at these models of doom and see if they justify so much confidence.

First, the blurb which opens Ehrlich's book is clearly wrong. My copy was published in 1969, a year in which about 53 million people died from all causes. It takes four seconds to read the blurb, so for one person to die each second, 31.5 million—about 60 percent of all deaths—would have had to have been from starvation.

Taking the UN cause-of-death statistics and being as fair as possible by including as "starvation" any cause of death related to nutrition—diphtheria, typhus, parasitic diseases, etc.—we get about a million, or some 5½ percent. Dr. Ehrlich is off by a factor of 10.

Actually, world agriculture is keeping up with population—so far. At the Mexico City meeting of the AAAS in 1973, Dr. H. A. B. Parpia, the senior professional of the UN's Food and Agricultural Organization, told me that just about every country *raises* enough food to be self-sufficient. It's grown, but sometimes not harvested; or if harvested, not eaten. In many countries vermin get more of the crop than the people; insects outeat people almost everywhere. The pity is that the technology to harvest and preserve enough for everyone exists right now.

Now this anti-doom essay is not a Pollyanna exercise. There's no excuse for relaxing and saying hunger is a myth. It isn't. But a simple thing like mylar sheeting to line traditional grain-storage pits and keep out insects could stop famine in 20 percent of the world. Other simple technologies could prevent hunger elsewhere.

So we know how to do it. But we won't do it unless we're willing to try. We won't get anywhere sitting around crying "Doom!"

Yet according to Dr. Ehrlich's book, "The battle to feed all of humanity is over. In the 1970's the world will undergo famines—hundreds of millions of people are going to starve to death in spite of any crash programs embarked upon now."

The other side of the coin was expressed in the Hudson Institute's *The Year 2000*, which points out that the level of rice-yield per acre under cultivation in India has not yet equaled what the Japanese could do in the 12th Century. Another analyst, Colin Clark, has shown that if the Indian farmer could only reach the production levels of the South Italian peasant, there would be no danger of starvation in India for a good time to come.

In other words, it doesn't take miracle rice, fertilizers, and a high-energy civilization to hold off disaster in the developing countries. It only takes adding technology to traditional peasant skills: showing people how to use mylar and non-persistent fungicides for food storage along with peasant production methods long known in Asia.

Actually very simple measures can have a profound effect. If India were to suffer an invasion of monsters who deliberately killed every third cow, the remaining cattle would be healthier and yield a great increase in milk and cheese proteins available for humans. More proteins in childhood would cut back infant diseases like *kwashiorkor* and "red baby"; those diseases have the effect of permanently lowering adult I.Q. by about 20 points. What would happen if the next generation of a developing country were "20 I.Q. points more intelligent"?

But some doomsters will now quote Malthus: if we help those people feed themselves, they'll only breed to famine levels again. Some will add, "So what's the point of it?"

The best answer is that historically, people haven't done that. When nations reach a high level of technology—and of infant survival—the fertility rate falls. The U.S. appeared to be an exception to that with the "baby boom" of World War II, but now that squiggle in the fertility rate is passed; the girls born in 1944 are over 30 years old now, and the number of girls born per fertile woman in the U.S. has fallen to all-time lows.

There's another form of doom not so fashionably discussed: the Marching Morons (that is, the least successful) tend to have the most children. It's one we must face, but it's doubtful that before 2000 it will have destroyed our social institutions.

As a matter of fact, given present population trends, the U.S. won't have many more people in 2000 than now. Population is growing; there's a "bow wave" generated by the "baby boom"; but best projections show us peaking in about 2025 and population then declines to present level—where it stays.

Suppose that never happens, and we reach 350 million people before something stops the U.S. population growth. The area of the U.S. is about 9.5 million square kilometers; of that, some is water and some simply uninhabitable. Call it 8 million even, and we have a present population density of 25.4 people/km$^2$.

When we reach 350 million—and few projections show us getting there in 50 years—we'll have 43.5 people/km$^2$, a big increase. Some writers say that'll be sufficient to drive us all stark, staring mad. We'll be inundated with personal contacts, at each others' throats, sleeping in hallways, and generally miserable as civilization collapses.

Well, what civilized countries have population density higher than the doom-level, 43.5?

Practically all of them. West Germany, not an uncivilized place, has 244 people/km$^2$, equivalent to 1.9 *billion* people in the U.S.! Denmark has 114 people/km$^2$; France 93; England and Wales, 322. Even Scotland, with its highlands and islands and hills and moors, has 66.

What densities can people stand and remain sane? No one really has an answer to that. But the Netherlands, a charming place, has

319 people/km$^2$; the Channel Islands has 641; and Monaco, the densest place on Earth, has 16,000!

Of course the U.S. could not be packed like Monaco or England. We would not like it if our country were as thickly populated as Denmark (although our eastern seaboard is more densely populated in places right now); but surely we would not go insane if we lived as close together as the Scots!

Moreover, we have the technology right now to support a large population while preserving wilderness. Soleri's *Arcologies* is a fascinating book; he shows enormous cities built on a few square miles of land, leaving parks and woodlands between them.

Less ambitiously, Larry Niven and I have "designed" a city for a story about Los Angeles in the future. (It's called *Oath of Fealty*, and will be out late next year.) In our design, a 50-level building contains lodging, stores, conveniences, recreation, employment, and transportation for 250,000 people. The building is 2 miles on a side and sits on an area 4 miles on a side; 250,000 people in 16 square miles. Fewer than a hundred of these would hold the entire U.S. non-farm population—and the structure is not only small by Soleri's standards, but uses very little technology we don't already have.

When we began the story, incidentally, I thought it a bit farfetched that people might prefer to live in our "city" instead of the suburbs. Now, I've seen condominiums with full conveniences, recreation, and transportation; they cost more than the suburbs; yet most of their inhabitants are refugees from suburbia. It no longer seems fantastic at all. Why not live in a convenient place where you can walk to work, take an escalator to the opera, and a train to the beach?

No. The evidence is clear that the population bomb won't kill us or drive us mad within our lifetimes. Certainly we can't keep doubling populations as fast as we have in the past—but why assume we will? When Malthus made his gloomy predictions, someone running off the exponential growth equations would have calculated that England in 1970 would have 400 million people, instead of the present 55 million.

Population stability won't happen of itself; but most of the really alarming population growth has been through prolonging of life. Birth rates have declined through this century, but people live longer, despite wars, famines, pollution, insecticides, crowding, and all the other forms of doom. Since there's a limit to just how long anyone can live, the death-rate is due for a climb before 2000. Already many countries have aging populations; including the U.S., of course. It was never true that "over half the people are under 25" and it gets less true all the time. Much of the "population explosion" is a one-time artifact, and you can't simply apply equations of exponential growth to the 20th Century to predict the future.

The MIT models of doom use precisely three parameters to predict the world population, and take no real account of the difference between population growth among developed nations and developing countries.

Certainly population pressure can finish us off; but must we believe we'll get to the *Soylent Green* stage before something is done about it? The evidence is that the technologically-advanced countries have already done something about it; and certainly we won't be destroyed by overpopulation before 2020.

If we have defused, or at least delayed, the population bomb, what's the next thing to kill us? Pollution, usually. The MIT models indicate that we must limit capital expenditure, de-technologize before pollution does us in. Dr. Asimov says that if we survive going mad, we'll be up against it because of energy limits.

He's right, of course, and even more so when he points out that even if we're able to rip all the coal and oil out of the ground to set a match to it for heat, we would loose so much carbon dioxide that the greenhouse effect will raise the temperature of the Earth.

The temperature rise will either melt the icecaps—thus drowning the seacoasts—or (according to some climatologists) move so much water vapor over the poles—where it would freeze out—that we would start a new ice age. Either way it would not be pleasant.

That doom is only 100 years away. Few of us will see it, but our children might; and we'd be poor parents if we didn't worry about it. Some put industrial pollution as reaching killing levels far earlier, although almost all give us until 2000—25 years or so—to do something.

Yet pollution is easily conquered. We already have the technology to reduce any given pollution to any desired level. I have had a bottle of drinkable sewage—reclaimed—sitting on my desk. It only takes money and energy.

We can even do it without giving up essentials, although some luxuries such as electric can-openers, power carving-knives, heated swimming pools in individual backyards, perhaps even driving 400 miles to conventions instead of taking a train, might have to go. We certainly won't starve.

However, pollution control takes energy, specifically electricity; and electric generators are themselves polluting. This seems a dilemma with no way out.

Actually, it's artificial. We could right now be constructing fission plants to generate non-polluting electricity. Fission plants produce radioactive wastes that must be stored, and there's a small chance of a really bad nuclear accident, so they are not a feasible long-term answer; but we can build them, and don't only because of legal restrictions. Incidentally, we kill 50,000 people a year with automobiles and put up with it; what are the chances of that bad a nuclear accident each year? We also kill thousands to tens of thousands with emphysema and other consequences of pollution from our fossil-fuel plants; who weighs those real deaths against the theoretical ones from more nuclear power plants?

We will need fission, for a while; but it's a dead-end. It increases the heat loosed on the Earth (although not the $CO_2$); and fission plants require cold water for cooling, a resource we really are running out of fast.

Better we should burn hydrogen. Hydrogen ash is water; no $CO_2$ and no sulfur oxides. There will be oxides of nitrogen problems with any hot-fired boiler, so that eventually we'll have to get electricity from less efficient systems such as hydrogen fuel-cells working at low temperatures; but doom from electric plants burning hydrogen is a long time off.

Where to get the hydrogen? Hardly from fossil fuels, of course, and as to nuclear fusion, not only is it at least 30 years off (on a commercial scale anyway) but it isn't the great panacea to begin with. Not only is it likely to require cooling water, but fusion creates heat that wasn't on Earth before. We need energy that doesn't upset the planetary heat balance.

Fortunately, the Sun shines on tropical waters all over the Earth. The surface water in the tropics remains an even 20–25°C.; while the water in the depths below is about 5°C. That little temperature difference is equivalent to a waterfall of 90 feet.

We have the technology right now to generate electricity by using "hot" water on the tropical surface to boil something like propane or Freon, passing the low-temperature steam through a turbine, and condensing it on the other side with cold water brought it from the depths. No fuel needed.

The volumes of water per kilowatt-generated passing through the plant are similar to those already being pumped through conventional water-cooled fossil-fuel plants.

The Gulf Stream holds about 75 times as much energy as the U.S. now uses. The Sea of Cortez has somewhat more. It's all renewed by the Sun every day; you can't run out. There's enough energy in the tropics to run the world for a long time to come, and it doesn't pollute. In fact, bringing the cold water up from the bottom is the same as the natural phenomenon known as "upwelling"—and in areas of natural upwelling over half the world's fish are currently caught.

The deep cold water is full of nutrients; get them to the surface and sunlight, and you have plankton blooms. Shrimp and fish grow like mad. They can be harvested to help with the protein and food shortage.

As to how the energy gets from the sea-based generators in the tropics to the energy-hungry U.S.A., once the electricity is generated it can be used to hydrolyze water; and the resulting hydrogen can be pumped in the present natural-gas pipelines plus others like them we would build as needed.

This energy system is not just theory. It works. In 1929 Georges Claude, the inventor of the neon light, built an operating 20-kilowatt pilot plant outside Havana. That was nearly 50 years ago. Some analysts think the temperature-difference system is right now *economically* competitive with conventional coal-fired generators; and it takes no breakthroughs, unlike fusion.

My point, though, is that if one thing won't do it, something else will. This is the first generation in history to not only be concerned about ecology and conservation but also to have the resources to do something practical about them without condemning much of the world to starvation.

We live in one of the most exciting times in all history. Surely we can do better than cry doom!

# SURVIVAL WITH STYLE

Suddenly we're all going to die. Look around you: a spate of books, such as *The Doomsday Book, Eco-Doom,* and the like; and organizations such as "Friends of the Earth" and "Concerned Citizens" all say the same thing: Western civilization has been on an energy and resources spree, and it is time to call a halt.

The arguments are largely based on a book called *The Limits to Growth.* Written by a management expert for a group of industrialists calling themselves The Club of Rome, *Limits* may be the most influential book of this century. Its conclusions are based on a complex computer model of the world-system. The variables in the model are population, food production, industrialization, pollution, and consumption of non-renewable resources. The results of the computer study are grim and unambiguous: unless we adopt Zero-Growth and adopt it now, we are doomed.

The doom can take one of several forms, each less attractive than the others. In each case population rises, then falls drastically in a human die-off. "Quality of Life" falls hideously. Pollution rises exponentially.

Earth is a closed system, and we cannot continue to rape her as we have in the past; and if we do not learn restraint, we are finished. We have no alternative but Zero-Growth if we are to survive. One ZG advocate recently said, "We continue to hold out infinite human expectations in a finite world of finite resources. We continue to act as if what Daniel Bell calls 'the revolution of rising expectations' can be met when we all know they cannot."

Jay Forrester, whose MIT computer model is the main inspiration for Zero-Growth, goes much further. Birth control alone cannot do the job. It is clear from his model that only drastic reductions in health services, food supply, and industrialization can save the world-system from disaster.

Behind all those numbers there is a stark reality: millions in the developing countries shall remain in grinding poverty—forever.

And the West, under Zero-Growth, has only two choices: impoverishment through really massive sharing with the developing countries—which must, however, cease to develop; or to retain wealth while most of the world remains at the end of the abyss. Neither alternative is attractive, but there's nothing else we can do. Failure to adopt Zero-Growth is no more than selfishness, robbing our children for our own pleasures.

So say the computers.

I can't accept that. I want not only to survive, but to do it with style. I want to keep the good things of our high-energy technological civilization: stereo, rapid travel, easy communications, varied diet, plastic models, aspirin, freedom from toothache, science-fiction magazines, Selectric typewriters, Texas Instruments pocket computers, fanzines, fresh vegetables in mid-winter, lightweight backpack and sleeping bag—the myriad products that make our lives so much more varied than our grandfathers'.

Moreover, I want to feel right about it; I do not call it survival with style if we must remain no more than an island of wealth in the midst of a vast sea of eternal poverty and misery. Style, to me, means that nearly everyone on Earth should have hope of access to some of the benefits of technology and industry.

That's a tall order. The economists say it can't be filled. My wishes are admirable but irrelevant. Their models prove that.

I might accept their verdict if they had modeled the right system; but in my judgment they did not. They assumed that we live on

Earth. If that were true—that Earth was a closed system, the only place or planet available to us—then Zero-Growth might be the best of a number of unpleasant alternatives. But suppose it isn't. Suppose the economists have left something out of their models....

Arthur Clarke once said that when a grey-bearded scientist says something is possible, believe him; when he says that it's impossible, he's very likely wrong. That, I think, is as true in this case as anywhere else. When the economists, those propounders of "the dismal science," tell us that we are doomed, it's time to take a fresh look at the problem.

Forrester's models are basically ready to kill us through lack of food, lack of non-renewable resources, and pollution. If we can lick *those* problems we're all right. Oh, sure: there's obviously a finite limit to the number of people the Earth can support. I know how to manipulate exponential curves as well as anyone, and if we project population growth mindlessly ahead we come soon to the point at which the entire mass of the universe is converted into human flesh. So what? It isn't going to happen; population growth always declines with increasing wealth.

But there are powerful religions, whose adherents control large portions of the globe, which condemn birth control.

Well, yes. And I'm no theologian. But I cannot believe that any rational interpretation of scripture commands us to breed until we literally have no place to sit.

"So God created man in his own image, in the image of God created he him; male and female created he them. And God blessed them, and God said unto them, Be fruitful, and multiply, and replenish the earth, and subdue it; and have dominion over the fish of the sea, and over the fowl of the air, and over every living thing that moveth upon the earth."

I will leave theology to the theologians; but the command was, "Multiply and replenish the earth, and subdue it"; and surely there must come a time when that has been *done*? When there can be no doubt that we have been sufficiently fruitful? And surely dominion over the wild things of the earth does not mean that we are to exterminate and replace them? Surely even those of the deepest faith may without blasphemy wonder if we are not rapidly approaching a time when we shall indeed have replenished and subdued the earth?

I cannot believe that we will continue to breed until we have destroyed the world; and frankly, I think of no more certain way to insure that the developing countries continue to increase in population than to condemn them to eternal poverty through Zero-Growth. So let's leave the bogeyman of unlimited population expansion. We have the technology to limit family size when, inevitably, there comes the time when everyone, no matter what his religious conviction, believes that the earth has been replenished and subdued.

Our next problem is food production. Surprisingly, it's nowhere near as critical as is generally supposed. Now whoa! Please don't write me about all the starving people in the world: I do know something about the situation. I've also interviewed senior officials of the UN's Food and Agricultural Organization. There are very few countries that could not, over a ten-year average period, raise enough food to give their populations more than enough to eat.

The catch is the "over a ten-year period" part. The *average* crop production is sufficient; but drought, flood, and other natural disasters can produce famine through crop failures over a one-, two-, or three-year period. You see, there's no technology for storing the surplus. The West has known for a long time about seven fat years followed by seven lean years, but it took us centuries to come up with reliable ways to meet the problems of famine.

Our solutions have been two-fold: storage of food, and weaving the entire West into a single area through efficient transportation.

Drought-stricken farmers in Kansas can be fed wheat from Washington, beef from the Argentine, and lettuce from California.

But this takes industrial technology on a large scale. Even providing mylar linings for traditional dung-smeared grain storage pits in Africa is a high-technology enterprise.

Next, we waste hell out of land. Let's look at a few numbers. A hard-working person needs about 7000 "large" Calories a day, or $7 \times 10^6$ gram-calories. The sun delivers 1.97 calories per square centimeter per minute onto the Earth. Say about 10% of that gets through the atmosphere, and that the sun shines about 5 hours (300 minutes) per day on the average. Further assume that our crops are about 1% efficient in converting sunlight to edible energy. Simple multiplication shows that a patch 35 meters on a side will feed a man—about a quarter of an acre.

Okay, I'm being unfair. But I'm not all that far off; you should see what my greenhouse, 2.5 meters on a side, can produce in hydroponics tanks; and there's no energy wasted in distribution of the food. I do use electricity to run the pumps, but I'm lazy; handwork would do it.

The joker, of course, is that I use chemical nutrients that take a lot of energy to manufacture. My greenhouse is made of aluminum tubing and mylar plastic with nylon reinforcements. All high-technology items, as are the fungicides I use, and even the water-testing kit that lets me balance pH in the nutrients.

Hmmm. We're back to industrialization again. Now it's true enough that if the average Indian farmer could manage the productivity level reached by the Japanese peasant of the 12th Century, India would have no food problems; but it's not likely he'll get there without industrial help; and meanwhile the Japanese have had to move far ahead of their 12th Century output levels.

But it should be obvious that sufficient levels of industrialization and technology will overcome the food production problem for a long time to come. To get ridiculous about it, if 1% of New York City were covered with greenhouses, they would feed about 10% of the city's population; greenhouses on 1% of Los Angeles would feed ⅓ of its population. Clearly food production *per se* isn't going to be a limit to growth for a good long time; food production will be limited by an enforced halt in industrialization and technology.

So now we come to the binding point. Our bottleneck is the penalties associated with industrialization. If we can industrialize without polluting ourselves to death, or without running out of non-renewable resources, then we can all get rich; we can have survival with style.

But how can we do that?

In a series of articles in *Analog*, my friend G. Harry Stine described what he called "The Third Industrial Revolution." Astute readers may even have noticed similarities between Harry's articles and my stories; as indeed they should, for Harry's articles were one of the most exciting events of my life.

Oh, sure; intellectually I knew that we could do all sorts of marvelous works in space; but Harry brought it home to me. His articles gave the *feel* of space industrial operations. In my judgment his phrase "Third Industrial Revolution" should become as standard a term as "industrial revolution" is now. *Should*; and I hope will; but it's not inevitable. The Zero-Growth movement may strangle the Third Industrial Revolution in its cradle.

Anyway, I'm pleased to say that in 1975 Putnam brought out *The Third Industrial Revolution*, by G. Harry Stine, and I recommend it to every reader who's at all concerned about the future.

Harry argues that when it's steam-engine time, there will be steam engines; and when it's space-industry time there will be space industries. There I disagree; space operations are so capital-intensive—

that is, require such enormous initial investments—that they're different from either steam engines or computers. By their nature, space industrial operations require *access* to space; and access is not available to the backyard inventor, or even the fabulously rich eccentric. If I invent a better mousetrap, I can find an investor rich enough to build it; but for space industry there's no firm or consortium of firms that can come up with the initial investment. *If* private enterprise ever gets access to space, the game's over; then we'll get the Third Industrial Revolution whether we want it or not; but if the anti-technology chaps have their way, the Shuttle will be turned in new buses for transporting children across town, school lunches, higher welfare payments, compulsory seat belts in automobiles, subsidies for tobacco growers, and public campaigns to "fight nuclear pollution."

Let's assume that somehow we get to space, though; how does that help us with industrialization? How will that enable us to survive with style?

From here on, while I will keep them simple, and work it so that you don't have to follow them to understand my conclusions, I going to have to use some mathematics; in particular I must introduce a way to measure and speak precisely about energy.

The basic energy measurement is the erg. It's an incredibly small unit; about the amount of energy used up when a mosquito jumps off the bridge of your nose. In order to deal with meaningful quantities of energy, we have to use powers-of-ten notation. Example: $10^2 = 100$; $10^4 = 10000$; and $10^{28}$ is 1 followed by 28 zeros.

Table 2 is included to give some feel for the numbers.

We're concerned about non-renewable resources and pollution, right? Let's go to space and solve both problems.

Probably the worst offender in both categories is metal production; give us enough iron and steel, copper, aluminum, zinc, and

Table 2: Energies of Various Events

| Event | Ergs | Event | Ergs |
|---|---|---|---|
| **Mosquito taking flight** | **1** | Annual output, total U.S. installed electric power system, 1969 | $5.4 \times 10^{25}$ |
| Man climbing one stair | $10^{9}$ | World electric power produced, 1969 | $1.6 \times 10^{26}$ |
| Man doing one day's work | $2.5 \times 10^{14}$ | Thera explosion (largest single energy event in human history) | $10^{26}$ |
| One ton of TNT exploding | $4.2 \times 10^{16}$ | **Present annual global energy use** | $\mathbf{10^{29}}$ |
| U.S. *per capita* energy use, 1957 | $2.4 \times 10^{18}$ | Solar flare | $10^{31}$ |
| **Conversion of one gram hydrogen to helium (fusion)** | $\mathbf{6.4 \times 10^{18}}$ | Annual solar output | $2 \times 10^{39}$ |
| Saturn V rocket | $10^{22}$ | Nova | $10^{44}$ |
| **One megaton (TNT)** | $\mathbf{4.2 \times 10^{22}}$ | Exploding galaxy | $10^{58}$ |
| Total energy annual use, Roman Empire | $10^{24}$ | Quasar, lifetime output | $10^{61}$ |
| Krakatoa | $10^{25}$ | Big Bang | $10^{80}$ |

**Energy;** *Little Bug to Big Bang*—One erg is the energy required to accelerate one gram at one centimeter per second per second over a distance of one centimeter.

lead, and surely we'll have our problems licked. After all, it's mine tailings that produce some of the really horrible pollution; copper refineries that poison so many streams; and those belching steel mills that make Pittsburgh a sight to behold (if you can see it through the smoke); and it's processing all those metals that really burns up the energy.

Give us metals free and clear, and the rest is easy. Give us enough metals and we'll industrialize the world. Besides, if we can do *that* in space, we can probably do anything else that has to be done. Consequently, I'll use metal production as my illustrative example.

In 1967—the last year I have complete figures for—the United States produced 315 million tons of iron, steel, rolled iron, aluminum, copper, zinc, and lead. (I added up all the numbers in the almanac to get the figure.) It works out to $2.866 \times 10^{14}$ grams of metal. Assume we must work with 3% rich ore, and we have $9.6 \times 10^{15}$ grams of ore to work with, or 10.5 billion tons.

It sure sounds like a lot. To get some feel for the magnitude of the problem, let's put it all together into one big pile. Assuming our ore weighs about 3.5 grams per cubic centimeter, we have $2.73 \times 10^{15}$ $cm^3$, or a block $1.39 \times 10^5$ cm on a side. That's a block less than 1.5 kilometers on a side; something more than a cubic kilometer, less than a cubic mile. Or, if you like it as a spherical rock, it's less than two kilometers in diameter.

There are something more than 300,000 rocks that size in the asteroids, and 3% ore isn't too bad a guess at their composition. Hmmm.

But we're dealing with the world, not the U.S.A., so let's give the whole world the per capita metal production of the United States; since we export a good bit of ours anyway, surely that's enough. So we take our 315 million tons and multiply by 2.2 billion, and divide by 200 million, to get *3.465 billion tons* of pure metal, $1.05 \times 10^{17}$ grams of 3% ore. That's $3 \times 10^{16}$ $cm^3$, or a rock 4 kilometers in diameter. There are well over 100,000 of *those* out in the Belt.

Well, we won't run out of metals. Only, of course, we have to process those metals.

For a moment forget they're out there in the Belt and imagine our rock is now in orbit around Earth. We want to get the metals out of it. Let's assume we do it by brute force. We're going to boil the whole rock.

It takes about 2000 calories per gram to boil iron. That's about the worst case for us, so we'll imagine our rock is entirely iron for the moment. It's going to take a lot of energy: $8.8 \times 10^{27}$ ergs, to be exact. That's something like twenty thousand one-megaton hydrogen bombs. Where'll we get the energy?

Hmm. The sun delivers at Earth orbit 1.37 million ergs a second per square centimeter, and out in space we can catch that with mirrors. There are 31 million seconds in a year, or $4.32 \times 10^{13}$ ergs/$cm^2$-year out there. We need $2 \times 10^{14}$ $cm^2$ of mirror, or one big round one 80 kilometers radius. Too big; even in zero-g that would be unwieldy. But a hundred of them 8 km radius doesn't sound so bad, or even a thousand at 1.6 km radius.

Of course my mirrors aren't going to be 100% effective—but then I'm not going to boil the whole blooming rock either. Nor do I seriously propose that we bring in the entire world metal supply from space, or that all the metal is simply consumed with no recycling. I'm looking for ballpark figures.

Note, by the way, that there's been absolutely no pollution on Earth so far. All the waste is out there where it can't hurt us. But we've still got problems, of course. After all, my metals are *not* in Earth orbit; they're out there in the Belt and they've got to be moved here—and that's going to take *energy*.

So let's see what it does take. To get from Ceres to Earth you've got to have a change in velocity—that's delta-v—of about 7 kilometers a second. By definition energy is mass given a velocity change, so we can quickly figure out how much energy we need by the formula $KE = ½mv^2$, and come up with $2.45 \times 10^{11}$ ergs for every gram moved into Earth orbit from Ceres. We're going to move our whole rock, all $10^{17}$ grams of it, so we'll need $2.6 \times 10^{28}$ ergs just to move it; about 10% of the world's present annual energy budget; not excessive in return for our entire metal supply.

But $10^{28}$ ergs is a lot of energy, and we're far away from the sun; I can't use sunlight for *that*. (Maybe I can, but we'll rule it out.)

So I need 61,000 one-megaton hydrogen bombs, which is quite a few; best I find another way. Note that if I don't find another way I may yet use the bombs; we needn't worry about radiation and fallout and the like out in space. My bombs are nothing compared to what the solar wind is doing during a flare. But how else might I work the transportation problem?

I need an invention: hydrogen fusion, which gives me, if I've got an efficient reaction, $6.4 \times 10^{18}$ ergs per gram of hydrogen "burned." I'm unlikely to have 100% efficiency, so you can multiply the number I come up with by whatever factor makes you happiest. If you think my system is 10% efficient, try that. Elsewhere I've described one kind of space-drive that will work given fusion.[2]

So I apply my fusion-ion drive, and discover I need to fuse $4 \times 10^{9}$ grams of hydrogen, which sounds like a lot, but it's really only 4000 metric tons, not so very much after all; quite a small ship could carry it. It's the amount of hydrogen in something like a cubic kilometer of water once I've thrown away the oxygen (which surely isn't polluting!) and we've got a lot of cubic kilometers of water on Earth.

I'll also need to get my hydrogen out to Ceres from Earth, which requires $9 \times 10^{21}$ ergs, less than a Saturn rocket can deliver, or the energy obtained in fusing another 1.5 kilograms of hydrogen. Clearly we're not going to run out of hydrogen for a long, long time. Even if we must go to deuterium—"heavy" hydrogen—we won't run short; and recall, there's likely to be some ice out there in space. We may not need to do all our hydrogen processing here on Earth.

So. For the price of one to a few thousand cubic kilometers of water each year, I've brought home all the metals we need to give the entire population of Earth as much metal as the U.S. produced in 1969. If we do nothing else in space—if we come up with no startlingly new processes as described by Harry Stine in his fascinating book—we'll have licked pollution and dwindling resources, thereby letting the developing countries industrialize, and thereby whipping the food production crisis for a while.

---

[2] See "Life Among the Asteroids," Galaxy, 1975 July.

Sure: there's a limit to growth. But with all of space to play with, I'll be happy to leave the problem for my descendants of 10,000 years hence to worry about.

I can hear the critics sputtering now. "But-but-but—what does this madman think he's *doing*? Flinging numbers like that around! It's absurd!"

Really? Remember, we fling quantities like that here on Earth right now; and I've after all assumed that we're going to supply the whole world with metals at the rate we produce them from all sources—including recycling—here at ground level U.S. of A. What's so absurd about it? Oh, no, we won't be operating on this scale for a few years; but then we weren't producing all those tons of steel back in 1930 either. Even Forrester's worst crunch model doesn't finish us off before 2020—a year in which we might very well be able to move asteroids around, boil them up for processing, and bring the resulting metals down for use on Earth. There's almost exactly as much time between now and 1930 as now and 2020.

Yes, we live on a finite Earth; but there's a whole solar system out there, just waiting for use to use it. We've only to lift our heads out of the muck to find not only survival, but survival with style.

# BLUEPRINT FOR SURVIVAL

This may be a unique century in many ways. In one respect it certainly is: this is the first time that mankind has had the resources to leave Earth and make his home in the solar system. No one doubts that we can do it. It takes only determination and investment.

Alas, we may be unique in another way: ours may be the only century in all of history when mankind can break free of Earth. Our opportunity may not come again, *per omnia secula seculonim*. Thus it could be that we have it in our power to condemn our descendants to imprisonment forever.

I've written about survival with style: how we can, if we will, usher in the Third Industrial Revolution through exploitation of space, and thereby supply Earth with non-polluting energy and metals for millennia. One reader commented as follows: "I remain skeptical. By the time man is forced to accept population control, the world is going to be in a sadder state than it is now. And I doubt if nations will give up their armaments and their free school lunches in order to get the resources to mine the asteroids until the situation is so bad that we probably can't mine the asteroids in time to save us."

Unfortunately he may be right. There is no end to foreseeable crises, and enough of them could so deplete our resource base and technological ability that when we realize that we *must* go to space, we won't be able to get there. Furthermore, anti-technological sentiments are no joke; a great number of influential intellectuals have embraced Zero-Growth, condemn technology, and seem to want the next generation to atone for the sins of our forefathers. They do not appear to want themselves to atone; I haven't seem many leading intellectuals giving up their own luxuries, much less necessities, in order to make amends for the "rape of the Earth," "eco-doom," or the rest of what engineers and technologists are accused of. *We* shall continue to enjoy; but after us, The Deluge. Our children shall pay.

And of course if Zero-Growth has its way, our children *will* pay; but ours won't pay as much as the children of the people in the developing countries. These kids are doomed with no chance at all.

Do not misunderstand. Were Earth our only source of energy and resources I should probably myself be crying Doom. As it is, I fully support many conservation measures—and in fact I was writing pro-conservation articles as early as 1957. I've no use for wasters of Earth's bounty. But I've less use for those who would condemn most of the world to eternal poverty when there are ways that we can do something about it.

Incidentally, the Club of Rome, which sponsored the original computer studies leading to *The Limits to Growth* and provides much of the intellectual fuel for Zero-Growth, has sponsored a second report entitled *Mankind at the Turning Point* (MATTP). This book, unlike *Limits*, is supposed to hold out some hope for the poor. By looking at the world as a set of 10 "regions" we can, say the authors of MATTP, divide the wealth and sustain what they call "balanced growth."

Unfortunately they never tell us how. As one reviewer put it, "I do not find any clear explanation of the ways in which balancing out the regions of the world would lead to any lessening of the total demands of human civilization on the planet's living-space, resources, and vital eco-systems." (Frank Hopkins, in the October 1975 *Futurist.*) Moreover, the MATTP plan demands foreign aid at the rate of some $500 billion *a year* at the end of a 50-year development period. True, there are plans with less massive foreign aid donations; but all are truly enormous, and like Zero-Growth must be started *now* or we are all doomed.

And this is nonsense. No politician is going to run for office on a platform of international bounty. No democratic—or communist—nation is going to shell out limited wealth at that rate. And even if, by some miracle, the western nations were to divvy up with everyone else, the Second Report can't challenge one feature of *The Limits to Growth*: no matter how wildly successful we are in imposing Zero-Growth and population control, in 400 years the game will be over. We will have run out of non-renewable resources. Mankind will have no choice but to give up high-energy civilization and return to some kind of pastoral society.

Surely this is not a desirable goal? There may be those who dream of the simple life (and a lesser number who will actually choose to live it), but surely only a madman would impose it on everyone else without dire necessity? If there is any alternative, must we not take it?

There are alternatives. They aren't even very expensive compared to the MATTP plan. Take, for example, the detailed plans of Princeton professor Gerard K. O'Neill.

Details of what have come to be called "O'Neill colonies" were first widely published in the September 1974 issue of *Physics Today*. The plan has been modified somewhat since that time—by a week-long NASA-sponsored conference of some of the biggest names in space exploration, for instance—but the basic concept remains the same: building self-sustaining colonies in space. O'Neill colonies have a major advantage: they are not only self-sustaining, but will be capable of building *more* colonies without further investment from Earth. When the first ones have been completed, Earth need pay no more. In addition, the colonies will be able to make important contributions to Earth's economy.

There's been a great deal of excitement in the science community, and of course among science-fiction fans, although oddly enough most SF writers haven't put much about O'Neill colonies in print.

In my own case I assumed others would, and I was waiting for new details. Even so, much of the SF community is aware of the O'Neill concept. "Life in Space" has become a regular program item at science-fiction conventions.

The basic O'Neill plan is for colonies able to support from 10 to 50 thousand people each. They will be located in the L4 and L5 points of the Earth-Moon system. Since not all readers know what that means, and the location is important to the economics of the project, let me take a moment to explain.

The equations of gravitational attraction are so complex that we can't really predict where planets, satellites, moons, etc., will be after long periods of time. Given high-speed computers we can make approximations, but we can't precisely solve problems involving three or more bodies except in special cases. In 1772 Lagrange discovered one of those special cases, namely, that when a system consists of three objects, one extremely large with respect to the rest, and another very small with respect to the other two, there are five points of stability: that is, things that get to those points tend to stay there. These are often called "Lagrangian points" and are designated as L1 through L5. They are illustrated in the figure opposite.

Three of the five are not really stable: that is, if an object is perturbed out of L1, L2, or L3, it won't tend to return. The other two, L4 and L5, are dynamically stable; left to themselves things put there will stay forever.

Points L4 and L5 are called *Trojan points*, because in the Sol-Jupiter system these points are occupied by a number of asteroids named after Trojan War heroes. The Trojans trail Jupiter, while the Greeks lead. Unfortunately the custom of naming the eastern group for Greeks and the western for Trojans wasn't established before one asteroid in each cluster was named for the wrong class of hero; thus there's a Trojan spy in the Greek camp, and vice versa.

Because of perturbing influences of other planets, Trojan points aren't really "points"; the Trojan asteroids drift around within a sausage-shaped area about one AU (93,000,000 miles) in diameter, while objects in the Earth-Moon Trojan points would tend to drift a few thousand miles one way or another. No matter; they're stable enough. Colonies and supplies, once they arrive at L4 or L5, won't go anywhere. The points are, of course, 240,000 miles from Earth and an equal distance from the Moon.

O'Neill colonies will be big. Even the first model, which is intended as an assembly base and factory, will be several kilometers in diameter. Later models will be larger. One design calls for a cylinder six kilometers in diameter and several times that in length, with "windows" running lengthwise to let in sunlight, large mirrors outside to focus more sunlight, and everything from farms and houses to trout streams in the "land" areas under the windows. The cylinders slowly rotate to provide artificial gravity. The exact gravity wanted isn't known yet, but it will certainly be less than that of Earth, possibly low enough that man-powered flight (yes, I mean people with artificial wings) will be not only feasible, but the usual means of personal transportation. As O'Neill points out, a great number of energy-consuming activities required for civilization on Earth can be greatly simplified in the colonies.

It's possible to wax poetic about the idyllic life in O'Neill colonies, but I won't do that. In the first place, I may be far-out technologically, but I don't think people are likely to live in Utopian style no matter how pleasant their environment. The important point is that life can be pleasant, and certainly possible, in space colonies.

These colonies are to be self-sufficient: they have more than enough agricultural area to feed their inhabitants. They are self-generating, with a duplication time of under ten years; over the long haul they could be built fast enough to accomodate some of Earth's surplus population. That, however, is not a major selling point, and we'll ignore it here.

Most importantly for our purposes the O'Neill colonies can sell power to Earth. It is perfectly feasible to collect solar radia-

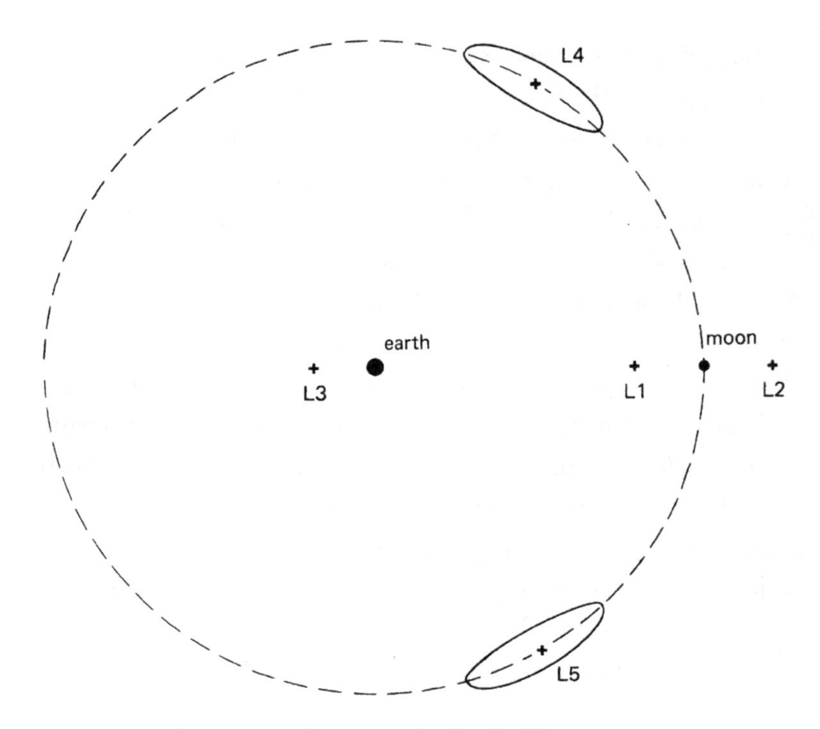


*Lagrangian Points in the Earth–Moon system*

tion, convert it to electricity, and beam the juice down to Earth by microwave. Tests show that cycle, from DC to DC, to be about 65% efficient—and of course most of the wasted energy doesn't get to Earth in the first place. There are a number of designs for the Earth-based receivers. The one I like best is a grid of wires several meters above ground; energy densities underneath are low enough to let cattle graze in the pastures below the grid.

All this sounds lovely, but surely it's a bit far-fetched? No. O'Neill colonies use present technology. There are no super-strong materials and no magic systems. We could now begin building an O'Neill colony this year, occupy it in 1990, and by the year 2000 have a couple more of them built. In which case we could also be supplying about as much power to Earth as the Alaskan pipeline will. In 20 more years space could supply nearly all U.S. electric power.[3]

So why don't we do it?

It's bloody expensive, that's why. Make no mistake: this would be a costly undertaking, on a level of effort compared to the Interstate Highway System, or the Vietnam War. It would not, in my judgment, be nearly so expensive as Zero-Growth, but unfortunately the costs of space colonies are *visible*. They're direct expenditures. The costs of Zero-Growth are hidden, since the most costly part is in potential not used and goods not created.

In the December 5, 1975 *Science* (the prestigious publication of the American Association for the Advancement of Science) Dr. O Neill presents an economic analysis of satellite solar-power stations (SSPS's) and space manufacturing facilities. He comes up with total costs ranging from a low of $31 billion—about the proportion of GNP that Apollo cost—to a high of $185 billion. He also discusses benefits from the electric power produced by SSPS's, and concludes that over a 40-year period the facilities would show actual profits from sales of power alone.

As one of the discoverers of the Law of Costs and Schedules ("Everything takes longer and costs more."), I tend to distrust Dr. O'Neill's numbers. It hardly makes any difference. The important point is that the program is feasible. We could afford it. Take a worst-case. Suppose it takes 25 years, and the total cost is 50 Apollo programs, that is, a round one trillion bucks. The money must be spent at $40 billion a year for the next 25 years, which comes to $200 a year for every man, woman, and child in the U.S. In my own family it would be about $1000 a year.

That's a lot of money. Worth it, I think; the benefits are literally

[3] Readers with more interest in O'Neill colonies might write the L5 Society, 1620 N. Park, Tucson AZ 85719. L5 publishes a newsletter and lobbies for NASA support for space colonies. Dues are $20 annually.

incalculable. For example, by the year 2000 the U.S. will need 2 billion tons of coal annually simply to operate our electric power system. Nuclear power plants could reduce that substantially, but the nuclear industry is in deep legal—not technological—trouble. It would be worth a lot to me simply to avoid the strip mines that 2 billion tons a year will require.

Moreover, the space budget isn't going to be simply tacked onto the national budget. All of the money will be spent here on Earth—people living in Lunar and space colonies have no need for Earth dollars, and what they physically import is tiny compared to the salaries that will be paid to Earth workers manufacturing products for the colony program. With $40 billion a year in high-technology industries, we can eliminate a number of "pump-priming" expenditures and dismantle several welfare and unemployment compensation schemes as well.

Of course we won't really need to spend that kind of money, and I suspect we can start getting returns on that investment before 25 years. O'Neill himself thinks in terms of some $5 billion a year, which works out to $25 a head for each person in the U.S.; and the colonies have got to be worth *that* if only in entertainment value.

Now how can something as complex as space colonies be built for that low a price? And wouldn't it be cheaper to build space manufacturing facilities in near-Earth orbit rather than going out to L5?

That's the beauty of the O'Neill concept. All the building materials for the colonies must, of course, be put into orbit—but it need not come from Earth. Most of the raw materials for the L5 colony will come from the Moon.

The Moon has one-twentieth the gravity well that Earth does. The colonies will be in stable Trojan points. Put those two data together and you reach an interesting conclusion: much of the mass of the colonies need never have been launched by rockets at all.

There are several devices for getting Lunar materials to the L5 point. One involves a simple centrifugal arm: a big solar-powered gizmo similar to the thing used to pitch baseballs for batting practice. It flings gup, such as unrefined Lunar ore (25–35% metal, from our random samples) out to the L5 point, and the law of gravity keeps it there. Refining takes place at the colony, and the slag is useful as dirt, cosmic ray shielding, and just plain mass. There's also oxygen in them there rocks.

Another workable device is the linear accelerator—a long electric sled as used in countless science-fiction stories. Both devices can be built with present technology.

Obviously, then, O'Neill colonies have a prerequisite, namely, a permanent Lunar colony. Now that's certainly within present-day technology; I once did studies that demonstrated that with technology available in the 60's we could keep astronauts and scientists alive for years on the Lunar surface, and things have come a long way since then.

The Lunar colony will need at least one near-Earth manned space station, since Earth-orbit to Lunar-orbit is the most effective way to transport large masses of materials from here to there. The Lunar shuttle will be assembled in space, and won't have all that waste structure that would enable it to withstand planetary gravity; thus it can carry far more payload per trip.

It's here that I think the profits come in. Skylab demonstrated that space manufacturing operations have fantastic potential profits. There are things we can make in space that simply cannot be made on Earth. Materials research benefits alone might pay for the space station. Certainly the potential for Earthwatch operations, pollution monitoring, better weather prediction, increased communications, and all the other benefits we've already got from space, will contribute to profits as well.

And once space shows a visible return on investment, we may well be on our way.

So. The prerequisite for the space station is the Shuttle; and there's the weak point. The Shuttle is in trouble. There are a number of Congresscritters who'd like nothing better than to convert the Shuttle into benefits for their own districts. There are plenty of intellectuals who continually cry "Why must we waste money in space when there are so many needs on Earth?" The obvious reply, that most of our expenditures on Earth seem to have vanished with no visible benefit, while our space program has already just about paid for itself in better weather prediction alone, does not impress them.

There are also the Zero-Growth theorists who see investment in space not as a mere waste of money, but as a positive evil.

We are close to breakthrough. For a whack of a lot less than we spend on liquor, or on cigarettes and cosmetics, on new highways we don't need, on countless tiny drains that fritter away the hopes of mankind, the United States alone could break out of Earth's prison and send men to space. The effect on future generations is literally incalculable. We *can* do it; but will we?

I wish I were sure that we would; or that if we of the U.S. don't do it, somebody else will; but I am not. There are just too many disaster scenarios. A Great Depression. War. The triumph of anti-technological ideology. The continued ruin of our educational system—in California, with 30 state universities, there is not one in which bonehead English is not the largest single class, and the retreat from excellence (called democratization and equality of opportunity) races onward. Any of these, or all of them at once, could throw away an opportunity that may never again come to mankind.

So what do we do?

For one thing, we can organize at least as well as the opposition. Science-fiction readers may have mixed emotions about "ecology" movements, consumerism, Zero-Growth, and the like, but I think we have not lost our sense of wonder, nor abandoned our hopes.

We have not given up the vision of man's vast future among the stars. We have not traded the future of man for a few luxuries in our time.

Unfortunately, we have no voice, or rather, we have a myriad of voices, none very effective. *That* at least we can remedy. There is a blanket organization whose goals I think most of us can accept, and I urge all of you to consider joining it. It is called the National Space Institute (NSI). Its first president was the late Wernher von Braun and the directors are professionals. Its purpose is to keep the faith; to keep alive the technology we need, to feed the dream, and ceaselessly to tell public officials just how important space is to all of us.

NSI dues are $15 a year, $9 for students. You may join by sending the money to National Space Institute, 1911 N. Fort Myer Drive, Suite 408, Arlington VA 22209. Dues and other contributions are tax-deductible. It has publications and such like, but that's not the reason to join. NSI's real benefit to members is as spokesman for our dreams.

In the 50's a number of us in the aircraft industry used to bootleg space research. There wasn't any budget for that crazy Buck Rogers stuff. Most of us believed we would see the day when the first man set foot on the Moon. We didn't believe we'd see the last one. I hope we haven't.

Like many of us who recall pre-Sputnik days, I alternate between hope and depression. Recently I have seen one hopeful sign, although it is a bit frightening.

It appears that the Soviets have built lasers sufficiently powerful to blind our infrared observation satellites. These satellites are in very high orbits, meaning that the Soviet lasers must be extremely powerful. One old friend who has remained in the industry told me at a New Year's party that the Soviets must be at least 5 years ahead of us, and this in a field in which we thought we were supreme.

Why is this hopeful? Isn't it rather frightening?

It's frightening if you think the Soviet Union may fall under or be under the control of convinced ideologists willing to trade part of their country for all of the world. There is nothing in Marxist ideology to forbid that—indeed, any communist who has the opportunity to eliminate the West and thus bring about the world revolution, and who fails to do it because of the price in human lives, is guilty of bourgeois sentimentality. So yes, it's frightening that the Soviets may have taken several long strides toward laser defense against ICBM's.

It's hopeful, though, in that it may stimulate us to get moving in large laser R&D. In my judgment, defense technology is the ideal way to conduct an arms race, if you must have an arms race. (And it takes only one party to start a race, unfortunately.) Defense systems don't threaten the opponent's civilian population. They merely complicate offensive operations, hopefully to the point where no sane person would launch an attack; and they give some hope that part of your own civilian population may survive if worst comes to worst.

If we can't justify space operations in terms of benefits to mankind, then perhaps we can sell them as defense systems? Big lasers can be used as space launching systems. If built they can put a good bit of material into orbit, thus making the manned space factory economically feasible and nearly inevitable; and once in Earth orbit, you're halfway to anywhere.

Specifically, we'd be halfway to an era of plenty without pollution; halfway to assuring that our descendants won't curse our memory, for throwing away mankind's hope for the stars.

# WHAT'S IT LIKE OUT THERE?

I'm writing most of this in a hotel room in Toronto, which is a lovely city but no place to be if you're alone on a Sunday night, and especially no place to be if you're from California: jet lag keeps you from getting to sleep at a normal hour, and the Provincial Police keep you from finding an open tavern....

It has been an interesting day. I've just taken part in a Canadian TV program called *The Great Debate.* The issue was, "Resolved: space research is a waste of time and money." Anyone who doesn't know which side I took shouldn't be reading these articles. Anyone who believes I lost the debate hasn't been reading them very long.

I slaughtered the poor chap. It helps that my opponent, John Holt, who is a charming fellow with a distinguished record in education, chose such a silly proposition to defend. It is trivially easy to show that space research has pretty well paid for itself already. I chose, in fact, to assert a new proposition: that the space program is I lie most important activity, excluding religion, in human history.

They tape *The Great Debate* in bunches, and prior to my own I watched another: Max Lerner and Toynbee's successor at Cambridge debating the proposition that Western civilization is in a state of irreversible and imminent collapse. As I listened it came to me that their whole conversation was irrelevant. It was as if a pair of very distinguished and learned professors in Paris were debating the same subject in 1491, unaware that this Genoese nut was making application to the Queen of Spain for a small fleet....

And of course I said as much in my own debate, and added that very probably in Iceland a few centuries earlier someone had won in a debate of, "Resolved, the voyages of Leif Ericson are a waste of time and money." To which Mr. Holt replied that the New World was accessible to the average family, while space never would be—that space would be restricted to scientists, astronauts, and military officers, a chosen few. The general public would never be able to go. He didn't say why.

Now at first the New World was pretty well inaccessible to anyone who couldn't get Queen Isabella to hock the crown jewels, and space is in the same situation at present. But just as the Americas were soon open to workers, farmers, administrators, soldiers, adventurers (some qualified, some merely desperate, some sent as sentence of courts), space will, probably well within my own lifetime, be open to large numbers—at least if Mr. Holt doesn't get his way. The only real question is how and when.

The how is simple technology. Shuttles will help a lot. Eventually, I trust, there will come the laser launching systems I've described before, which can put up privately-owned capsules, the equivalent of the covered wagon. There will be O'Neill colonies—which Mr. Holt particularly hates; Lunar bases; asteroid mining and refineries; Mars colonies; possibly ice mining on Enceladus for the Mars-terraforming project; all these and more are in the cards and there's no reason to suppose they'll be restricted to superheroes.

The when is a little harder to predict, but in fifty years for certain. It didn't take that long to get colonies established in the New World.

So what's it going to be like to live out there?

Well, first let's take the O'Neill colony, which is a huge cylinder in space. NASA figures we could have the first one before the year 2000 if we wanted it, and there are good numbers to show that it would pay for itself within a few years after its establishment: it can sell power to Earth, as well as serve as a base for extensive space manufacturing—and there are plenty of things that you can manufacture *only* in space.

The colony will be quite large, say a cylinder three kilometers in diameter and 10 to 20 kilometers long. Windows run the length of it to let in sunlight. Under the windows is land, ordinary dirt, with hills, streams, buildings, and such like. The whole thing rotates to give artificial gravity. Let's suppose the medical people have determined that a tenth of an Earth gravity is sufficient for long-term health; that means our cylinder rotates at .026 radians a second, or .25 revolutions per minute.

A colonist standing on the ground and looking up through a window above will see the stars swinging past once each four minutes. He'll also see his neighbors' fields and houses hanging in space above his head, which can be disconcerting until he gets used to it, after which it won't seem any stranger than seeing mountains in the distance.

Life in the O'Neill colony may be a bit strange, but it has its compensations. If the colonist is a farmer, he'll never have to worry about the weather. There won't be any rain—he'll have to irrigate—but on the other hand there won't be floods, storms, or droughts (so long as the engineers keep the watermakers going).

He will be able to calculate *exactly* how many hours of daylight his crops will get for the entire growing season. The only weeds and insects he'll encounter will be those brought aboard by the ecology teams.

Actually, one suspects a few pests will come along as stowaways. Imagine the town meeting after the sparrows have got loose. One faction wants them left alone. They're cute. Another advocates shotguns. Still another abhors guns, but is willing to send to Earth lor a supply of sparrowhawks. After four hours of shouting the council sets the matter aside for another day….

Machinists and mill workers will find their work little different from Earth, except that everything weighs only 10% as much. For production runs the colony probably has computer-controlled lathes and milling machines, but for one-of-a-kind items the machinists will have to do the work. There will undoubtedly be lawyers and doctors and storekeepers and librarians and tailors, none of whose business lives will be all that different from what it would be on Earth.

But after working hours things get more exciting. No freeways; no cars. No subways, either. In 10% gravity the simplest means of transportation is to fly with artificial wings. There might not be any other form of transport besides walking. Why should there be? (Well, for heavy hauling you might want a few electric trucks, but surely there's no need for any individuals to own cars or trucks.)

If flying is the usual transport, grocery shopping will be like New York City, where you buy a few items a day as you need them, rather than like California where you buy bags and bags once a week and transport them in a car.

Flying also means that everyone in the colony is accessible to everyone else; every place is easily accessible to anyone wanting to get there. This can drastically change the sociology. Houses will probably have roofs, not to keep the rain out, but to keep the neighbors from looking in. The house need not be anything more than a visual screen: it doesn't have any weather to control.

What all this does to the colony's mores isn't really predictable. There's little privacy. Parents will know pretty well what their teenage kids are doing. Whether this will make premarital sex more or less common isn't obvious—at least not to me. It depends partly on geography, I suppose: will there be any secluded places, dark and cozy? Dark comes when the windows are closed for the night, of course; the Sun only sets when the colony wants it to. Daylight saving time is silly in an O'Neill colony, because if you want more daylight, you simply program the window blinds to give it.

It may be that parents won't care much where their children are. There won't be any dangers in the colony; one presumes that airlocks to the outside and the like are controlled against accidental use, and also that there won't be many incompetents in the community—at least not people *that* incompetent. There remains the problem of crime.

It's hard to imagine jails in a space colony, although I suppose they could be built. It's hard to imagine space muggers in the first place, or that the colonists would put up with them. They might be enslaved to the community. The cost of shipping an unwanted colonist back to Earth would be slightly colossal. On the other hand, the environment is fragile enough that you certainly don't want anyone wandering around harboring burning resentment against the colony—especially not if he has suicidal tendencies. It would be all too easy to take a number of others along in a spectacular suicide.

We can presume, then, that the environment is *safe*; free of most of the dangers we live with here on Earth. Now in England the custom of dinner parties grew up only after Sir Robert Peel invented police; prior to that no one in his right mind went *anywhere* after dark, and when you visited friends you stayed at least for the night. When the London Police made the streets comparatively safe it became possible to visit for the evening and go back home for the night. Such factors will affect the colony patterns of friendship too.

On the other hand, there are dangers that we don't worry about here. The most significant would be leaks. It would take a very large leak to affect the colony, of course. Small ones would be costly (air isn't cheap when it has to be taken to orbit) but easily repaired before anyone felt their effects. Still, it seems reasonable that there would be a few major airtight structures, shelters, into which the colonists could crowd in the event of a major break in the pressure hull.

An interesting life, with kids learning to fly at an early age. I suppose when a parent tells a teenager he's grounded, he'll mean that quite literally.

So what do you do in such a colony? Well, what do you do now? It's simple enough to sit at home and watch TV whether you're in New York or Earth orbit. Some recreations won't be possible. No backpacking trip through the wilderness. Probably no sailboating: no wind, even if there's a lake. There may be fishing, but certainly no hunting.

On the other hand, there'll be cultural activities not available on Earth. Flying, of course; real flying, not dangling from an oversized kite, but man's ancient dream of flying like a bird. Aerial acts will probably become an art form, possibly involving a large portion of I the colony population. There can also be aerial ballet, with and without wings. Up in the center of the cylinder there's no gravity. Zero-g areas are easily accessible.

Games can be strange. With that large radius and slow rotation rate, the colonists won't easily be able to tell the difference between their artificial spin gravity and the real thing—not, that is, until they begin throwing things. As soon as you throw something, say a baseball, you'll know you don't have normal gravity. The ball's trajectory will be strange, and it will depend on which direction you threw it in. You'll also be able to throw the ball a very long way, so far that baseball may require much larger teams to cover the huge playing field.

In fact, any projectile motion is affected. Obviously, in one-tenth gravity you can throw a ball (or a javelin or a wrestling opponent) ten times as far as you could on Earth. A javelin-throwing athlete who can manage 285 feet on Earth would get 2850 feet-over half a mile—in the O'Neill colony gravity. Broadjumpers would also do well.

However, there's a problem. When you loft a thrown object in centrifugal gravity, you increase the time of flight; and the ballistics becomes strange indeed, due to an effect called the Coriolis force. What happens is this: from the viewpoint of an observer inside the spinning object, the "gravity" is radial. Objects dropped tend to fly

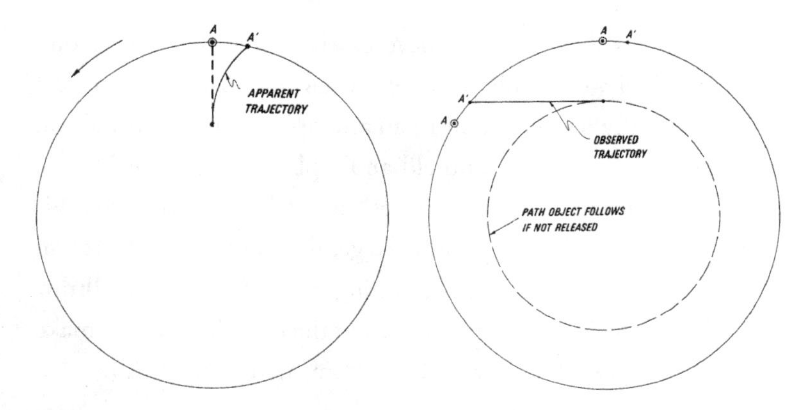


*Apparent Coriolis "Force"* *View of Non-rotating Observer*

*In these figures, A is the point directly "below" the object, and A' is the point on the hull that the object strikes when dropped. The separation depends on the rate of rotation and the height the object is dropped from. Figure 3 shows what someone in the colony would observe; figure 4, what would be seen by an outside observer.*

directly away from the center. They fall toward the "floor," and in 10% gravity as we have here, they fall rather slowly. It takes two full seconds for something to drop two meters.

While the object is falling, the "floor" is moving, so that the dropped object does *not* strike the spot directly under it. The discrepancy is related to the rate of spin and the radius of the spinning craft, and for something as large as an O'Neill colony you'd never notice it under normal circumstances; but if you throw the ball up, or loft it into an arcing trajectory, the effect can be *very* noticeable.

(I know: it isn't *really* that way at all. To an observer watching from outside there is no such thing as "centrifugal force," and the Coriolis effect I described in the last paragraph is also a pseudo-force. What happens is that the released object tends to fly along in a straight line tangent to the circle of motion; but the effect, as far as someone inside the colony is concerned, is as I described it. I've diagrammed the situation in figures 3 and 4.)

The result is that if we did have baseball in an O'Neill colony, the batted ball would follow an abnormal trajectory. The fielders could jump fifty feet in the air in an attempt to catch it. If the ball nevertheless falls into the outfield and a player snags it, he'll have to be careful not to aim his throw at the catcher. Exactly what his point of aim should be if he wishes to get the ball to home plate will depend on where the player is standing when he makes his throw. If the axis of the field is along the axis of the cylinder it could make quite a difference whether you threw from right or left field!

Conditions in a Lunar colony would be rather different. While it's only one-sixth Earth's, the gravity on the Moon is real, not artifical. Also, O'Neill colonies have to be built with a lot of open space. A Lunar base doesn't, and most models have the colony carved out of caves. It's certainly possible to roof over a large crater, and it will probably be done: but I doubt that there will be any large surface cities.

Lunar farmers have a problem. The Sun doesn't shine all the time. During the long Lunar night there's got to be heat and light for their plants. There are a lot of schemes to provide that, from full-time artificial light to mylar-roofed craters with an opaque roof that can be put on over it (and artificial lights, of course). You certainly have to cover any transparencies (large ones, anyway) during the night cycle. If you didn't, you'd lose all heat to radiation. The effective temperature of outer space is about -200°C (73 K) and heat radiates proportional to the fourth power of the temperature difference. Even here with Earth's atmosphere to catch some of that outgoing heat, it's always *much* colder on a clear night than a cloudy one—and in fact the Romans used the night sky to make ice cream in the Sahara. Maybe I'd better explain that.

Take one large pit, and fill it with straw. The idea is to insulate it as thoroughly as possible. Put a small container in the middle of the straw. At night you expose the pit to space. It radiates heat. In the day time you keep it covered with more straw and on top of it all place highly-polished shields or other reflective surfaces. Ice will form in a few days (provided that the night sky is clear, as it is in the desert). Enough for the Romans. Back to space.

Life on the Moon has been thoroughly described in science-fiction stories, and there's no point in my doing it again. For an excellent book on the uses of the Moon, see Neil Rusczic's *Where the Winds Sleep*. The Lunar colony is, after all, a complex cave with lower gravity than Earth's. (I hardly need mention Mr. Heinlein's classic *The Moon Is a Harsh Mistress*.)

*Zero* gravity is another story. Any long trip through space will have to be made by Hohmann transfer orbits, which use the lowest amounts of fuel, but which also take a lot of time: about a year and a half to get from Earth to Ceres, for example, and even a trip to Mars takes something over eight months.

It's possible to design ships so that they have artificial spin gravity of course. There are some problems with that, and since many readers would like to do some preliminary design work themselves, I'll give the equations here. Readers uninterested in the details can skip the next paragraphs.

Newton's First Law says that an object in inertial space wants to continue the same velocity (that's both direction and speed) forever. It takes a *force* to make any change in velocity. Gravity serves as the force to get moving objects into an orbit, exactly as the string serves to provide a force when you whirl a weight around on the end of a rope. In both cases the object wants always to go in a straight line, which is to say it wants always to go off in a direction tangent to its circle of motion. It does *not* "fly out from the center," although the result, as seen by an observer *moving with the system*, looks that way.

Thus if you stand on a moving carrousel it feels as if you're trying to fly out radially from the center, and in free space the "floor" of a centrifuge will be "down." If you let go of an object it will experience an acceleration relative to the carrousel, and for those inside the system that looks very much like gravity.

The acceleration is:

$$a_r = \omega^2 R \qquad \text{(Eq. 1)}$$

where $R$ is the radius of the rotation, and $\omega$ is the rate of rotation in *radians* per second. There are $2\pi$ radians in a circle, so if you multiply radians per second by 360 and divide by $2\pi$, you get degrees per second. Multiply the result by 60 and you have degrees per minute; divide the end result by 360 and you have revolutions per minute.

Going the other way,

$$\frac{\text{rpm} \times 2\pi}{60} = \text{radians/second} \qquad \text{(Eq. 2)}$$

Since force equals mass times acceleration (the most basic equation in Newtonian physics), it's easy to see that the force exerted by (and the tension on) the cord when you whirl a weight on a rope is:

$$F = ma = m\omega^2 R \qquad \text{(Eq. 3)}$$

where $m$ is the mass of the whirled object. This is the centripetal force, and it's real. If the cord were suddenly cut, the object would fly away in a straight line tangential to the circle of rotation. The velocity it would have is:

$$V_T = R\omega \qquad \text{(Eq. 4)}$$

and we're finished with the math.

Now we're ready to design a ship, and immediately we see the problem. The shorter the radius, the faster you have to spin the ship to get a given artificial gravity. Now it happens that the faster the spin, the worse the Coriolis effect. If the radius of rotation is long compared to, say, the height of a man, there's no big problem, but as it gets short there can be devastating physiological effects.

It seems silly enough now that we've put men into orbit, but at one time planners seriously thought space stations and ships needed something like a full Earth gravity to keep humans alive, and we did plans for such things. If you try for a full g in a ship of small radius, the Coriolis effect is so severe that a water-hammer is set up in the circulating system. A man could kill himself of stroke simply by turning his head rapidly in the wrong direction.

We now know that humans don't need a full gravity, and we suspect that a tenth might be enough forever. That can be arranged for a long trip if we send ships in multiples: join the ships with long cables and rotate them around each other. That's also very inefficient, of course: we have to duplicate life-support systems, etc. There's less dead weight in one large ship than in many small ones.

Also the tension in the cable can get quite high, as you can find from Eq. 3.

Maybe we don't need any gravity at all? True, the first Apollo astronauts came out much the worse for wear, and so did the first Skylab crew; but the interesting part is that the longer men stay in space, the better they adapt to it. The *Skylab Four* (third manned Skylab, in NASA's screwy counting system) crew came out in much better shape than did the second crew. Okay, in retrospect maybe it's not so surprising that the longer you stay in zero-g the better you adapt, but it did in fact surprise a number of space physiologists who had thought that a month of zero-g might be beyond human endurance.

A long trip in no gravity can be interesting. The accounts of the Skylab experience make for fascinating reading. They also show the need for experience in space. There were some terrible design faults in Skylab.

For instance: Skylab was the first space vehicle in which the astronauts ate at a table using spoons and forks, rather than squeezing everything from tubes and baggies. Their table was a mere pedestal that supported their food trays. There were seats, but those were seldom used: to stay in a sitting position in zero gravity requires that you bend at the waist and hold yourself bent. It puts a constant and severe strain on stomach muscles, and in fact those were the only muscles better developed when the crew landed than when they went up. The real problem, though, was the table itself.

It didn't do a very good job of holding the trays, to begin with. The tray lids were held down with what Lousma called "the most miserable latch that's ever been designed in the history of mankind or maybe before." Pogue said of the table, "I wouldn't want the people that designed that table to do anything else...."

Despite their attempt at normal meals, the Skylab astronauts never had much appetite. Part of that is due to less need for food: you're not working very hard in zero gravity. Also, the thinner air (kept at low pressure to avoid strain in the pressure bulkheads and such) doesn't transmit food smells very well.

Everyone had head congestion, caused by pooling of body liquids in the torso and head, so nothing tasted very good anyway.

However, they did eat.

With food in plastic bags (which were inside cans, which were supposed to be fitted into the trays on the table, but which often drifted loose because the cans didn't fit the trays very well) they could use spoons and forks. Eating in zero-g takes practice. You have to be careful to bring the spoon in a smooth arc from tray to mouth. Any hesitation and the food travels on in a straight line, probably into your eye.

The Skylab astronauts were almost constantly dehydrated, but never felt thirsty. The human organism is designed with a number of mechanisms to get the blood back out of the legs and up into the torso. So long as the legs are below the body those work fine; but when there's no such thing as "below," the blood gets into the torso and stays there. With all that fluid pooled in the abdominal region the thirst mechanisms don't work well, and the Skylab crews had to train themselves to take a quick drink every time they passed the water fountain. The fountain wasn't designed very well either, with metal nozzles that would have been easy to use on the ground, but which could chip teeth when not under control. The fountain buttons were so stiff that when the crewmen pushed them, the button didn't go down—the crewmen went up unless he was holding onto something.

Of course it's hard to blame the designers. Until Skylab nobody had any real experience at designing living quarters for space. Apollo was a ship, and there wasn't much room to move around in it. The crew mission was to get somewhere and come back, not live in space. Gemini was worse, and Mercury was downright primitive: when we stuffed people into the Mercury capsules they were fitted in precisely, without even room to straighten arms and legs.

John Glenn once said you don't ride a Mercury capsule, you wear it.

And prior to Mercury we hadn't any real experience at all. We flew transport planes in parabolic courses that might give as much as 30 seconds of almost-zero-g, and that was all we knew. I will not soon forget some of our early low-g experiments. Some genius wanted to know how a cat oriented: visual cues, or a gravity sensor? The obvious way to find out was to take a cat up in an airplane, fly the plane in a parabolic orbit, and observe the cat during the short period of zero-g.

It made sense. Maybe. It didn't make enough that anyone would authorize a large airplane for the experiment, so a camera was mounted in a small fighter (perhaps a T-bird; I forget), and the cat was carried along in the pilot's lap.

A movie was made of the whole run.

The film, I fear, doesn't tell us how a cat orients. It shows the pilot frantically trying to tear the cat off his arm, and the cat just as violently resisting. Eventually the cat was broken free and let go in mid-air, where it seemed magically (teleportation? or not really zero gravity in the plane? no one knows) to move, rapidly, straight back to the pilot, claws outstretched. This time there was no tearing it loose at all. The only thing I learned from the film is that cats (or this one, anyway) don't like zero gravity, and think human beings are the obvious point of stability to cling to....

Future dwellers in zero gravity won't have so much to worry about. The nine Skylab crewmen dictated hours and hours of notes on design improvement, this time not theory, but well-founded in experience. The next space station (if we get one) should be a lot more comfortable.

And life in zero gravity, the Skylab crew tells us, is fun. Almost no one simply went from one place to another. It was impossible to resist turning somersaults, flips, ballet twirls, just for the sheer hell of it. Most of us saw the TV demonstrations: waterballs floating in air, tiny planetary systems that could be set in motion by blowing gently on them. There were other lovely experiments, and just plain play, all described beautifully in a book I recommend, Henry S. F. Cooper's *A House in Space* (Holt, Rinehart and Winston, 1976).

I haven't yet mentioned the asteroids, which are different again. They have *some* gravity, but very little. Things do fall, but slooowly. On Ceres, for example, you can jump about 125 feet into the air (oops! into space) and it takes over a minute for the round trip. On very small rocks you can jump clean off, never to return.

There are dangers on intermediate sizes, too—ones too large to jump from. For example, some respectable asteroids—several kilometers in diameter—have such low gravity that if you jumped hard you'd not leave it forever, but it would take hours to go up and come back down again. You could easily run out of air.

And so forth. I've tried to describe some aspects of life in the asteroid belt in my stories "Tinker" and "Bind Your Sons to Exile," and other science-fiction writers have written hundreds of such. It will be interesting to see how well we've done: despite all the stories about zero-gravity (and a number of SF fans among the engineers who designed Skylab), there were a lot of surprises once we actually got up there.

There will be more.

# BLACK HOLES AND COSMIC CENSORSHIP

As I write this, California courts are trying to decide whether the police have the power to seize copies of the film *Deep Throat*, and my friend Earl Kemp may be headed for jail due to violation of censorship laws. Thus I'm tempted to write about censorship, but since this is a science column and not a political essay I don't suppose I'll be able to.

However, one should never underestimate the ingenuity of a columnist....

I suppose, though, I'd better stay with science and cosmology. I've just got the latest on gravitation research; that seems like a good safe topic. I mean, how far from censorship can you get?

I expect most readers are at least vaguely aware of the on-going research on detection of gravitational waves, but it won't hurt to summarize a bit. In the Newtonian universe, and in many other theories as well, gravity is a "force" that acts through a field. That is, these theories postulate that although it is $10^{40}$ times weaker than electromagnetism, it is not fundamentally different.

This essential similarity holds true in the realm of special relativity also. Special relativity, you'll recall, states that no material object and no signal can travel faster than light. There's a good bit of evidence for special relativity, and no really good counter-theory lurking in the wings to take its place.

The *general* theory of relativity, however, is another breed of cat entirely. There are several contenders in that realm, and experimental evidence offers no clear-cut way to choose one or the other. General relativity does away with gravity fields altogether. In that theory, gravity results from the geometry of space.

Whether gravity fields "exist" or merely result from geometry, theorists believe gravitational attraction propagates with the speed of light. Thus, if matter is created—or destroyed—the rest of the universe won't be instantly affected, but must wait until the gravitational effect, traveling at lightspeed, reaches it.

Thus "gravitation waves," which will have a frequency and an amplitude much like light, but which may also have some rather strange properties as well.

In theory, if we could detect and examine gravitational waves, we might be able to tell whether they result from a field and are thus similar to magnetism, or if they are merely a property of space and its geometry. Unfortunately, gravity is an incredibly weak force. It requires the mass of the whole Earth merely to pull things with a puny 980 cm/sec$^2$ acceleration—and we can overcome that with lather small magnets, or chemical rockets, or even our own muscles when we jump.

Because gravity is so weak, it's hard to play with. You can't turn on a "gravity wave generator" and fiddle with the resulting forces to see if they refract, or can be tuned, or whatever. You can't wiggle a mass to generate gravity waves, because you can't get a large enough mass held into place to be wiggled. It's not even possible to blow off an atomic weapon, turning some matter into energy, and measure the effect of the matter vanishing; the effect is just too small to be noticed, and it's hidden among the rather drastic side effects.

However, there are a number of theoretical ways that gravity waves might be generated by the universe: stars collapsing into black holes or neutronium would do it, for example. The universe might be riddled with gravitational waves, but they'd be terribly weak, and require delicate and sophisticated apparatus to detect them.

Some years ago, Dr. Joseph Weber of the University of Maryland decided to build a gravity wave antenna. He took a large aluminum cylinder and covered it with strain guages. The idea was that so long as the cylinder were acted on only by the steady gravity of earth, it would be in a stable configuration; but if a gravity wave passed through it, the cylinder would be distorted, and the strain guages would show it.

He had to compensate for temperature, and isolate it from vibration, and worry about a lot of other things, but the technology had been developed: the antenna was built. It was incredibly sensitive, able to detect distortions on the order of an atomic diameter. It was also able to detect student demonstrations outside the library, trucks rumbling along the highway a mile distant, and other unwanted events.

The solution to the latter problem was simple: build another copy of the antenna and place it 1000 kilometers away; now hook the two together, and pay no attention to an event that doesn't affect both. Such "coincidences" should be due to a force affecting both antennae, and since even earthquakes take time to propagate—and their effects move much slower than lightspeed—the output should be reliable.

Unfortunately, it isn't as straightforward as that. The instruments must be very sensitive, and thus there's a lot of chatter from them. By the laws of chance, some of this chatter will be simultaneous, or near enough so, and thus you are guaranteed some false positive results. The output of the gravity wave detectors, therefore, needs careful analysis to decide what's real data and what's chance.

Weber immediately got results. He got a lot of results, far too many for chance. Unfortunately, there were far too many for cosmologists to believe. As a result of Weber's early reports, some cosmologists estimated that as much as 98% of the universe must be inside black holes.

The argument went this way: something is producing gravity waves. We can't see enough matter to account for the events, but normal matter falling into a black hole would produce gravity waves. Therefore–

There were other cosmologists who wanted to believe this for different reasons. Readers familiar with black holes must excuse me: it's now necessary to discuss their basics for a moment.

A black hole is a theoretical construct that can be derived from both general relativity and the older Newtonian universe; in fact, the first speculations about black holes come from Laplace in 1798. If you take enough matter and squeeze it small enough, you will eventually get so much gravitational force that nothing can prevent the matter from continuing to collapse.

In Einsteinian terms, the space around the matter becomes curved into a closed figure, but the result is the same: the matter is squeezed to infinite density. Long before it reaches that state, though, there is a region around the matter at which the escape velocity is greater than the speed of light.

The effect of that should be pretty obvious. If light can't escape, you can't see down into the hole. Moreover, anything that goes down in the hole can never come out: that is, if you accept the speed of light as the top limiting velocity of the universe, nothing can come out, ever.

The area at which space is curved into a closed figure—or the region at which the escape velocity is equal to the speed of light—is known as an *event horizon*, and interestingly enough both Newtonian and Einsteinian equations give the same location to it.

It is the region at which:

$$R = \frac{2\,GM}{c^2} \qquad \text{(Eq. 5)}$$

where $R$ is the radius from the center, $G$ is the universal constant of gravitation, $M$ is the mass, and $c$ is the speed of light. For the Sun, that radius is on the order of three kilometers: if the Sun is ever squeezed that small, we'll never be able to see it again.

An observer diving into the black hole would never know when he had crossed the event horizon. He could continue to send signals to his friends outside, and as far as he could tell, they would go right on up and out.

Those outside the hole, though, can never under any circumstances receive information from inside it.

Now, as it happens, if we measure the total amount of matter in the universe, and plug that in for $M$ in Eq. 5; and we take the furthest object we can observe and plug that in for $R$; then the equation almost balances.

Almost, but not quite. There isn't enough matter in the universe; we're missing from 20% to 90% depending on whose figures you use for $M$ and $R$.

If the equation were to balance, space would be curved into a closed figure at the boundaries of the universe; and we'd live in a closed universe.

Eventually, in a closed universe, those galaxies receding from us will stop and come back, and the whole universe will be packed into a big wad at the center. What happens after that is debatable, but a number of cosmologists want badly to believe in a closed universe.

It also means, of course, that we live inside a black hole ourselves—that is, our whole universe *is* a black hole.

If we don't live in a closed universe, the receding galaxies will go right on receding, and this disturbs some theorists. Thus, Weber's coincidences were welcome in many cosmological circles. Others tried to build gravitational antennae to confirm his results.

Then a second startling result came out of Weber's shop. It appeared that there was a 12-hour sidereal cycle to the coincidences, and furthermore, that this cycle was related to the galactic plane. In other words, gravitational waves originated in the galactic center.

We have a good estimate of the distance to the galactic center, and thus were able to estimate how large an effect at the center of the galaxy would be required to deliver that much force to us out here on our spiral arm. The result was once again dismaying. Far too much energy was apparently being turned into gravitational waves.

Now the energy radiating from the galactic center could be either sprayed out in all directions, obeying the inverse-square laws, or it could be "beamed" into the galactic plane. Obviously less total energy is involved if it is "beamed," but what mechanism might account for that?

The speculations were many, imaginative, and varied; they were also rather frightening.

Let's take a moment to go back to black holes. When matter gets dense enough to satisfy Eq. 5, and the event horizon forms, things don't just stop there. The matter goes on collapsing; we just can't see it any longer.

In fact, *nothing* can stop the collapse. In theory, the matter should quite literally become infinite in density. Infinity is a troublesome concept: how can infinite density be present in a finite universe? The answer is obvious: in some respects, the matter no longer remains in the universe at all.

When gravitational forces have got to this point, we have what is known as a *singularity*: a point at which normal laws simply do not apply.

Actually, things are worse than that. Not only don't normal laws apply, but the relativity equations suggest that *no* laws apply. Strange things happen in the region of a singularity. Time is reversed. Conservation laws don't work. Causality is a joke: if you could get into the region of a singularity, you really could go back in time and assassinate your grandfather.

In fact, anything could happen, and science ceases to exist; and you don't even have to physically go to the singularity for this to take place. All you have to do is be able to observe one directly, and science has just gone down the drain. That bothers a lot of theorists and scientists, and rather disturbs me as well.

If there is a naked singularity—that is, a singularity not covered with an event horizon—then, at least in potential, there is no order to the universe.

Out of that thing might come ghosties and ghoulies and things that go bump in the night.

What, then, may we do to save science? Why, invoke censorship, of course. (I told you never to underestimate the ingenuity of a columnist.)

The kind of censorship invoked is called rather whimsically the "Law of Cosmic Censorship," which states that "There shall be no such thing as a naked singularity." *All singularities must be decently clothed with an event horizon.*

Given cosmic censorship, a number of interesting laws about black holes may be proved: that they never get smaller, that if one is rotating it can't be sped up until the escape velocity is smaller than the speed of light, and a number of other rules that are collectively known as the laws of black hole dynamics.

Unfortunately, cosmic censorship deprives science-fiction writers of some of their best stories. Hmm. Cosmic censorship is unfair to SFWA....

It does it this way. If all black holes are covered with event horizons, it follows that you can't plunge into a black hole and come out elsewhere or elsewhen. Actually, if you plunge into a random black hole, all that could ever come out anywhere would be a stream of undifferentiated subnuclear particles; for all their fantastic properties, singularities do retain one feature, namely that gravitation in their region is rather high, sufficient to disassociate not only the molecular, but the atomic, structure of anything visiting them.

On the other hand, if a star about to collapse into black hole status is rotating fast enough, some solutions to the Einstein tensor suggest that the singularity formed will be a doughnut; you could dive through that and come out in one piece, provided the doughnut were large enough.

Large enough means a galactic-sized black hole, I'm afraid; stellar-size black holes will still get you too close to the singularity so that you can't use them for transportation. Furthermore, what you come out to on the other side is not, according to the equations, our universe at all. What it will be like, no one can say, except that it will have in it a copy of the black hole you dove through to get there.

So, turn around and dive back, of course; but that doesn't work. You go through and out again, all right, but into a third universe different from either of the other two. The black hole is still there, so try again—and come out in a fourth, and there behind you is that rather tiresome black hole again.

Is any of this real, or are we playing with ideas? No one really knows, of course. The most we can say is that the people who can solve Einstein tensors come up with that kind of result.

It's rather discouraging for science-fiction writers. Here we thought we had a new way to get faster-than-light travel, what with black holes connecting us to another universe, or—just possibly—to another region of our own, and the very people who gave us the black holes go on to prove we can't use them.

Still, maybe there's a way out. Perhaps someone will find a solution. But they can't so long as the Law of Cosmic Censorship is enforced, because the singularities decently covered with event horizons can't come out and affect our universe.

Back to Weber and gravitational waves. One of the models constructed to account for the enormous gravitational energy generated in the center of the galaxy had a very large singularity lurking down there. Suns fell into it, and as they were eaten, gravity waves poured out. It was a rather depressing picture, our galaxy being eaten alive like that.

Then a number of other laboratories constructed gravitational antennae. Bell Laboratories, an English group, the Russians—all made gravity wave detectors. In each case their equipment was supposed to be an improvement on Weber's.

None of them found any coincidences at all. People began to wonder just what Weber had done, and to doubt his results.

At the Cambridge Conference of experimental relativists in the summer of 1974, though, the picture changed again.

The people who had built "improved" gravity wave antennae reported no results whatever.

Weber continued to report results, but with a change (I'll get back to it in a moment).

And two other groups—one at Frascatti, Italy; the other at Munich, Germany—had built carbon copies of Weber's antenna. They got coincidences. Whatever Weber was observing, others have independently observed something similar now.

Meanwhile, Weber did a re-analysis of his coincidences, using a computer rather than human judgment to define just what was a coincidence. The result was startling. He still gets events—but they are no longer concentrated in the galactic plane. The sidereal coincidences have gone away, and with them has gone the evidence for the large singularity eating the galaxy.

Moreover, Dr. Robert Forward, of Hughes Research at Malibu, California, has constructed his own gravity wave antenna. Since lasers were invented at Hughes Labs, it's no surprise that Forward's antenna employs them. He has three big weights at the apexes of a right triangle.

Lasers measure the precise distance of each weight from the others. A gravity wave will presumably distort that triangle, and thus be detected.

Forward has "events" too. They seem to coincide with the kinds of things Weber gets but, as I write this, no serious attempt has been made to compare results.

For that matter, the Munich people have just got started. They were quite surprised, by the way; they'd thought Weber's results were some kind of artifact.

It appears, then, that some kind of gravity waves do travel about through the universe; at least something that can affect large aluminum cylinders hundreds of kilometers apart is operating here.

The next step is to see if these events have any relationship to the bursts of x-ray energy detected by Vela satellites. At the moment that's not possible, and of course there are a lot more gravity wave events than x-ray events; but if the x-ray events are accompanied by coincidences on the gravity antenna, we'll know a lot more about both.

We may then be able to decide what gravity is: a force, or a distortion of geometry. We may be able to learn more about black holes, and what happens inside them, and who knows, those trips to alternate universes could be a real possibility.

Until we get rid of cosmic censorship, though, we'll never know what happens to the volunteers who go exploring down black holes.

# FUZZY BLACK HOLES HAVE NO HAIR

Black holes have no hair, but they're fuzzy. Because they're fuzzy, they're not really black.

Classical black hole theory dictates several laws of black hole dynamics. Some aren't too interesting, but the Second Law says that the area of the event horizon can never decrease, and increases as matter and energy are pumped into the hole. This means that black holes never get smaller. Feed them matter and/or energy and they grow.

That lets us deduce one thing instantly. What happens if a normal matter and an anti-matter black hole collide? Well, nothing that wouldn't happen if two normal matter holes (or two anti-matter holes) collided, of course. The holes eat each other to form one larger than either, but we'll never know which ones contain matter or anti-matter. In fact, the question is meaningless.

You see, black holes have no hair.

This is a convenient way to say that everything we'll ever know about a black hole can be deduced from three parameters. Once you specify the mass $M$, the angular momentum $J$, and the charge $Q$, you've said it all. Nothing remains but location, which isn't important for the physics of the hole, but may be for the physicist who wants to study it.

Mass we understand. It doesn't really matter whether that mass is in the form of energy or matter; Einstein's $e = Mc^2$ takes care of that, and down in the hole it's irrelevant whether the rest mass is $e$ or $M$.

Angular momentum comes from the rotation of the object before it collapsed. Naturally it's conserved, so that if the star were rotating, the thing inside the hole rotates as well. It also rotates *fast*, just as a skater speeds up in a spin when she pulls her arms in.

The last parameter, charge, is just what it says, and it gives us a way to move a black hole around. If it isn't charged, feed it charged particles until it is, then use magnets to tow it.

The laws of black hole dynamics say you can never recover the rest mass energy (that's the $Mc^2$ energy, of course) of the original body. It's lost forever. Even shoving anti-matter down the hole gains you nothing.

However, you can get energy out of a spinning black hole. Up to 29% of the rotational energy is available, and in the case of a star that's a *lot*. To get it you throw something down the hole, and one of the things that comes out is gravity waves.

In our experience gravity waves are puny things, but we're a long way from their source. Up close is another matter entirely. You could be torn apart by them—as the characters in my story, "He Fell Into a Dark Hole," very nearly were.

Most of what we know about black holes comes from Stephen Hawking of the University of Cambridge. Many physicists think Hawking is to Einstein what Einstein was to Newton, and he's still a young man. This year Hawking has added quantum mechanics to classical black hole theory, and he's ruined a lot of good science-fiction stories.

In 1973 Larry Niven and I went out to Hughes Research Laboratories in Malibu. The laser was invented at Hughes, so of course they do a lot of laser research there. They're also among the top people in ion drive engines, and they've done a lot with advanced communication concepts.

All that was fascinating, but we went to talk with Dr. Robert Forward, who's known as one of *the* experts on gravitation. I'd met him because he liked "He Fell Into a Dark Hole" and had been kind enough to call and tell me so.

Bob Forward is the inventor of the Forward Mass Detector, a widget that can track a tank miles away by mass alone. It can't distinguish between a tank at a mile and a fly on the end of the instrument, but if you use two and triangulate you're safe enough. His detector can also be lowered into oil wells, or towed behind an airplane to map mass concentrations below.

After lunch we talked about black holes. Dr. Forward was particularly interested in Stephen Hawking's then-new notion that tiny black holes might have been formed during the Big Bang of Creation. Since the Second Law predicts that they never get smaller, there should be holes of all sizes left. Some might be in our solar system.

They would come to rest in the interior of large masses. There might be quite a large one inside the Sun, for example, and even in the Earth and Moon as well. A very large mass hole, say $10^8$ kilograms, would still be very small: about $10^{-19}$ centimeters radius. An atomic radius is around $10^{-9}$ centimeters, very large compared to such a hole, so that the hole couldn't eat many atoms a day, and wouldn't grow fast.

Black holes inside the Earth or Sun aren't too useful because they're hard to get at. Bob Forward wanted to go to the asteroids. You search for a rock that weighs far too much for its size. Push the rock aside and there in the orbit where the asteroid used to be you'll find a little black hole.

You could do a lot with such a hole. For example, you could wiggle it with magnetic fields to produce gravity waves at precise frequencies. There might be all sizes of holes, even down to a kilogram or two.

It sounded marvelous. Larry and I figured there were a dozen stories there. I'd already written my black hole story, and Larry hadn't, so he beat me into print with a thing called "The Hole Man." All I got from the trip was a couple of articles and columns.

Well, Larry's story was reprinted in his collection *A Hole in Space*, while the columns I did about little black holes have been forgotten—I hope!

I'm glad I have nothing in print about tiny black holes, because Hawking has just proved they can't exist. Oh, they can be formed all right, but they won't be around very long. It seems that black holes aren't really black. They radiate, and left to themselves they get smaller all the time. The Second Law needs modifying.

Stephen Hawking's new paper was submitted to the 1974 Gravity Research Foundation prize essay contest. The GRF was founded by Edison's friend, stock market analyst Roger Babson. It's been around for many years, and received scornful treatment in Martin Gardner's *Fads and Fallacies in the Name of Science.* It may or may not have deserved that in the 50's, but for a number of years the leading people in gravitational theory have been entering the competition.

Hawking won first prize from the GRF in 1971 with his paper on cosmic censorship and black hole dynamics. This year he took only third prize, first going to a Cal Berkeley astronomer. Even third prize was enough to tear Larry's "The Hole Man" to shreds. (Not that Hawking ever mentioned science fiction; but then the Pioneer probes weren't intended to wreck all our stories about Jupiter, either.)

Hawking points out that Einstein's general relativity, which produces most of the primary equations for black holes, is a classical theory. It doesn't take quantum effects into account.

Hawking corrects this. In quantum theory a length, $L$, is not fixed. It has an uncertainty or fluctuation on the order of $L_0/L$, where $L_0$ is the Planck length $10^{-33}$ cm.

Since there is uncertainty in the length scale, it follows that the event horizon of the black hole isn't actually fixed. It fluctuates through the uncertainty region.

In fact, the black hole is *fuzzy*, and energy and radiation can tunnel out of the hole to escape forever. It's the same kind of effect as observed in tunnel diodes, where particles appear on the other side of a potential barrier.

Since black holes have no hair, although they do have fuzz, the quantum radiation temperature—that is, the rate at which they radiate—must depend entirely on mass, angular momentum, and charge.

It does, but I'm not going to prove it to you. Hawking uses math that I *can* tool up to follow, but I'm not really keen on Hermetian scalar fields, and I doubt many readers are either. If you want his proof, send a dollar to Gravity Research Foundation, 58 Middle Street, Gloucester MA 01930 and request a copy of Hawking's paper "Black Holes Aren't Black."

Hawking shows that the temperature of a black hole is

$$T = \frac{10^{26}}{M} \text{ Kelvin} \qquad \text{(Eq. 6)}$$

where $M$ is mass in grams, and the lifetime of a black hole in seconds is

$$t_L = \frac{M^3}{10^{28}} \text{ seconds} \qquad \text{(Eq. 7)}$$

Using my Texas Instruments SR-50 that handles scientific notation and takes powers and roots in milliseconds, it wasn't hard to work up table 3 from these equations.

There are more numbers than we need, of course. It's a consequence of the pocket computer. Not long ago I d have had to use logs and slide rule, and I'd have done no more than I needed. Now look.

The first thing to see is that small holes have uninteresting lifetimes. In order for one to be around long enough to use it, the hole must be massive.

Table 3: Lifetimes of Black Holes

| Description | Mass | Radius | Lifetime |
|---|---|---|---|
| Kilogram | $10^3$ gm | $1.48 \times 10^{-25}$ cm | $10^{-19}$ sec |
| Billion gm | $10^9$ | $1.48 \times 10^{-19}$ | 0.1 |
| 2375 tons (U.S.) | $2.15 \times 10^9$ | $3.19 \times 10^{-19}$ | 1 |
| — | $6.81 \times 10^{11}$ | $1.01 \times 10^{-16}$ | 1 year |
| — | $6.81 \times 10^{13}$ | $1.01 \times 10^{-14}$ | $10^6$ |
| — | $1.47 \times 10^{15}$ | $2.18 \times 10^{-13}$ | $10^{10}$ |
| Ceres | $8 \times 10^{23}$ | $1.2 \times 10^{-4}$ | $10^{36}$ |
| Earth | $5.98 \times 10^{27}$ | $8.88 \times 10^{-1}$ | eternal |
| Sun | $1.99 \times 10^{33}$ | $2.96 \times 10^5 = 3$ km | |
| Galaxy | $10^{11}$ suns | $2.96 \times 10^{16} = .03$ light-year | |
| Universe | $10^{22}$ suns | $2.96 \times 10^{27} = 3 \times 10^9$ ly | |

Any black holes formed in the Big Bang would be $10^{10}$ years old now; so if they weren't larger than a small asteroid they're gone already. Worse, that exponential decay rate defeats us even if we find a hole just decayed to an interesting size. It will still vanish too fast for use.

So there went Larry's story, and two I had plotted but hadn't written, and I suspect a lot of other science fiction as well. Sometimes I feel a bit like Alice when she protested, "Things *flow* here so!"

But it's what we get for living in interesting times, and it ought to teach my friend Larry not to rush into print ahead of me....

# BUILDING *THE MOTE IN GOD'S EYE*

with Larry Niven

Collaborations are unnatural. The writer is a jealous god. He builds his universe without interference. He resents the carping of mentally deficient critics and the editor's capricious demands for revisions. Let two writers try to make one universe, and their defenses get in the way.

But. Our fields of expertise matched each the other's blind spots, unnaturally well. There were books neither of us could write alone. We had to try it.

At first we were too polite, too reluctant to criticize each other's work. That may have saved us from killing each other early on, but it left flaws that had to be torn out of the book later.

We had to build the worlds. From Motie physiognomy we had to build Motie technology and history and life styles. Niven had to be coached in the basic history of Pournelle's thousand-year-old interstellar culture.

It took us three years. At the end we had a novel of 245,000 words… which was too long. We cut it to 170,000—to the reader's great benefit. We cut 20,000 words off the beginning, including in one lump our first couple of months of work: a prologue, a battle between spacegoing warcraft, and a prison camp scene. All of the crucial information had to be embedded in later sections.

We give that prologue here. When the Moties and the Empire and the star systems and their technologies and philosophies had become one interrelated whole, this is how it looked from New Caledonia system. We called it

## MOTELIGHT

Last night at this time he had gone out to look at the stars. Instead a glare of white light like an exploding sun had met him at the door, and when he could see again a flaming mushroom was rising from the cornfields at the edge of the black hemisphere roofing the University. Then had come sound, rumbling, rolling across the fields to shake the house.

Alice had run out in terror, desperate to have her worst fears confirmed, crying, "What are you learning that's worth getting us all killed?"

He'd dismissed her question as typical of an astronomer's wife, but in fact he was learning nothing. The main telescope controls were erratic, and nothing could be done, for the telescope itself was on New Scotland's tiny moon. These nights interplanetary space rippled with the strange lights of war, and the atmosphere glowed with ionization from shock waves, beamed radiation, fusion explosions.... He had gone back inside without answering.

Now, late in the evening of New Scotland's 27-hour day, Thaddeus Potter, Ph.D., strolled out into the night air.

It was a good night for seeing. Interplanetary war could play hell with the seeing; but tonight the bombardment from New Ireland had ceased. The Imperial Navy had won a victory.

Potter had paid no attention to the newscasts; still, he appreciated the victory's effects. Perhaps tonight the war wouldn't interfere with his work. He walked thirty paces forward and turned just where the roof of his house wouldn't block the Coal Sack. It was a sight he never tired of.

The Coal Sack was a nebular mass of gas and dust, small as such things go—eight to ten parsecs thick—but dense, and close enough to New Caledonia to block a quarter of the sky. Earth lay somewhere on the other side of it, and so did the Imperial Capital, Sparta, both forever invisible. The Coal Sack hid most of the Empire, but it made a fine velvet backdrop for two close, brilliant stars.

And one of them had changed drastically.

Potter's face changed too. His eyes bugged. His lantern jaw hung loose on its hinges. Stupidly he stared at the sky as if seeing it for the first time.

Then, abruptly, he ran into the house.

Alice came into the bedroom as he was phoning Edwards. "What's happened?" she cried. "Have they pierced the shield?"

"No," Potter snapped over his shoulder. Then, grudgingly, "Something's happened to the Mote."

"Oh for God's sake!" She was genuinely angry, Potter saw. *All that fuss about a star, with civilization falling around our ears!* But Alice had no love of the stars.

Edwards answered. On the screen he showed naked from the waist up, his long curly hair a tangled bird's nest. "Who the hell–? Thad. I might have known. Thad, do you know what time it is?"

"Yes. Go outside," Potter ordered. "Have a look at the Mote."

"The Mote? *The Mote?*"

"Yes. It's gone nova!" Potter shouted. Edwards growled, then sudden comprehension struck. He left the screen without hanging up. Potter reached out to dial the bedroom window transparent. And it was still there.

Even without the Coal Sack for backdrop Murcheson's Eye would be the brightest object in the sky. At its rising the Coal Sack resembled the silhouette of a hooded man, head and shoulders; and the off-centered red supergiant became a watchful, malevolent eye. The University itself had begun as an observatory founded to study the supergiant.

This eye had a mote: a yellow dwarf companion, smaller and dimmer, and uninteresting. The Universe held plenty of yellow dwarfs.

But tonight the Mote was a brilliant blue-green point. It was almost as bright as Murcheson's Eye itself, and it burned with a purer light. Murcheson's Eye was white with a strong red tinge; but the Mote was blue-green with no compromise, impossibly green.

Edwards came back to the phone. "Thad, that's no nova. It's like nothing ever recorded. Thad, we've got to get to the observatory!"

"I know. I'll meet you there."

"I want to do spectroscopy on it."

"All right."

"God, I hope the seeing holds! Do you think we'll be able to get through today?"

"If you hang up, we'll find out sooner."

"What? Och, aye." Edwards hung up.

The bombardment started as Potter was boarding his bike. There was a hot streak of light like a *very* large shooting star; and it didn't burn out, but reached all the way to the horizon. Stratospheric clouds formed and vanished, outlining the shock wave. Light glared on the horizon, then faded gradually.

"Damn," muttered Potter, with feeling. He started the motor. The war was no concern of his, except that he no longer had New Irish students. He even missed some of them. There was one chap from Cohane who....

A cluster of stars streaked down in exploding fireworks. Something burned like a new star overhead. The falling stars winked out, but the other light went on and on, changing colors rapidly, even while the shock wave clouds dissipated. Then the night became clear, and Potter saw that it was on the moon.

What could New Ireland be shooting at on New Scotland's moon?

Potter understood then. "You bastards!" he screamed at the sky. "You lousy *traitor* bastards!"

The single light reddened.

He stormed around the side of Edwards' house shouting, "The traitors bombed the main telescope! Did you *see* it? All our work—oh."

He had forgotten Edwards' backyard telescope.

It had cost him plenty, and it was very good, although it weighed only four kilograms. It was portable—"Especially," Edwards used to say, "when compared with the main telescope."

He had bought it because the fourth attempt at grinding his own mirror produced another cracked disk and an ultimatum from his now dead wife concerning Number 200 Carbo grains tracked onto her New-Life carpets....

Now Edwards moved away from the eyepiece saying, "Nothing much to see there." He was right. There were no features. Potter saw only a uniform aquamarine field.

"But have a look at this," said Edwards. "Move back a bit...."

He set beneath the eyepiece a large sheet of white paper, then a wedge of clear quartz.

The prism spread a fan-shaped rainbow across the paper. But the rainbow was almost too dim to see, vanishing beside a single line of aquamarine; and that line *blazed*.

"One line," said Potter. "Monochromatic?"

"I told you yon was no nova."

"Too right it wasn't. But what is it? Laser light? It has to be artificial! Lord, what a technology they've built!"

"Och, come now." Edwards interrupted the monologue. "I doubt yon's artificial at all. Too intense." His voice was cheerful. "We're seeing something new. Somehow yon Mote is generating coherent light."

"I don't believe it."

Edwards looked annoyed. After all, it was *his* telescope. "What think you, then? Some booby calling for help? If they were that powerful, they would send a ship. A ship would come thirty-five years sooner!"

"But there's no tramline from the Mote to New Caledonia! Not even theoretically possible. Only link to the Mote has to start inside the Eye. Murcheson looked for it, you know, but he never found it. The Mote's alone out there."

"Och, then how could there be a colony?" Edwards demanded in triumph. "Be reasonable, Thad! We hae a new natural phenomenon, something new in stellar process."

"But if someone *is* calling–"

"Let's hope not. We could no help them. We couldn't reach them, even if we knew the links! There's no starship in the New Cal system, and there's no likely to be until the war's over." Edwards looked up at the sky. The moon was a small, irregular half-disk; and a circular crater still burned red in the dark half.

A brilliant violet streak flamed high overhead. The violet light grew more intense and flared white, then vanished. A warship had died out there.

"Ah, well," Edwards said. His voice softened. "If someone's calling he picked a hell of a time for it. But at least we can search for modulations. If the beam is no modulated, you'll admit there's nobody there, will you not?"

"Of course," said Potter.

In 2862 there were no starships behind the Coal Sack. On the other side, around Crucis and the Capital, a tiny fleet still rode the force paths between the stars to the worlds Sparta controlled. There were fewer loyal ships and worlds each year.

The summer of 2862 was lean for New Scotland. Day after day a few men crept outside the black dome that defended the city; but they always returned at night. Few saw the rising of the Coal Sack.

It climbed weirdly, its resemblance to a shrouded human silhouette marred by the festive two-colored eye. The Mote burned as brightly as Murcheson's Eye now. But who would listen to Potter and Edwards and their crazy tales about the Mote? The night sky was a battlefield, dangerous to look upon.

The war was not really fought for the Empire now. In the New Caledonia system the war continued because it would not end. Loyalist and Rebel were meaningless terms; but it hardly mattered while bombs and wrecked ships fell from the skies.

Henry Morrissey was still head of the University Astronomy Department. He tried to talk Potter and Edwards into returning to the protection of the Langston Field. His only success was that Potter sent his wife and two sons back with Morrissey. Edwards had no living dependents, and both refused to budge.

Morrissey was willing to believe that something had happened to the Mote, but not that it was visible to the naked eye. Potter was known for his monomaniacal enthusiasms.

The Department could supply them with equipment. It was makeshift, but it should have done the job. There was laser light coming from the Mote. It came with terrific force, and must have required terrific power, and enormous sophistication to build that power. No one would build such a thing except to send a message.

And there was no message. The beam was not modulated. It did not change color, or blink off and on, or change in intensity. It was a steady, beautifully pure, terribly intense beam of coherent light.

Potter watched to see if it might change silhouette, staring for hours into the telescope. Edwards was no help at all. He alternated between polite gloating at having proved his point, and impolite words muttered as he tried to investigate the new "stellar process" with inadequate equipment. The only thing they agreed on was the need to publish their observations, and the impossibility of doing so.

One night a missile exploded against the edge of the black dome. The Langston Field protecting University City could only absorb so much energy before radiating inward, vaporizing the town, and it took time to dissipate the hellish fury poured into it. Frantic engineers worked to radiate away the shield energy before the generators melted to slag.

They succeeded, but there was a burn-through: a generator left yellow-hot and runny. A relay snapped open, and New Caledonia stood undefended against a hostile sky. Before the Navy could restore the Field a million people had watched the rising of the Coal Sack.

"I came to apologize," Morrissey told Potter the next morning. "Something damned strange *has* happened to the Mote. What have you got?"

He listened to Potter and Edwards, and he stopped their fight. Now that they had an audience they almost came to blows. Morrissey promised them more equipment and retreated under the restored shield. He had been an astronomer in his time. Somehow he got them what they needed.

Weeks became months. The war continued, wearing New Scotland down, exhausting her resources. Potter and Edwards worked on, learning nothing, fighting with each other and screaming curses at the New Irish traitors.

They might as well have stayed under the shield. The Mote produced coherent light of amazing purity. Four months after it began the light jumped in intensity and stayed that way. Five months later it jumped again.

It jumped once more, four months later, but Potter and Edwards didn't see it. That was the night a ship from New Ireland fell from the sky, its shield blazing violet with friction. It was low when the shield overloaded and collapsed, releasing stored energy in one ferocious blast.

Gammas and photons washed across the plains beyond the city, and Potter and Edwards were carried into the University hospital by worried students. Potter died three days later. Edwards walked for the rest of his life with a backpack attached to his shoulders: a portable life-support system.

It was 2870 on every world where clocks still ran when the miracle came to New Scotland.

An interstellar trading ship, long converted for war and recently damaged, fell into the system with her Langston Field intact and her hold filled with torpedoes. She was killed in the final battle, but the insurrection on New Ireland died as well. Now all the New Caledonia system was loyal to the Empire; and the Empire no longer existed.

The University came out from under the shield. Some had forgotten that the Mote had once been a small yellow-white point. Most didn't care. There was a world to be tamed, and that world had been bare rock terraformed in the first place. The fragile imported biosphere was nearly destroyed, and it took all their ingenuity and work to keep New Scotland inhabitable.

They succeeded because they had to. There were no ships to take survivors anywhere else. The Yards had been destroyed in the war, and there would be no more interstellar craft. They were alone behind the Coal Sack.

The Mote continued to grow brighter as the years passed. Soon it was more brilliant than the Eye; but there were no astronomers on New Scotland to care. In 2891 the Coal Sack was a black silhouette of a hooded man. It had one terribly bright blue-green eye, with a red fleck in it.

One night at the rising of the Coal Sack, a farmer named Howard Grote Littlemead was struck with inspiration. It came to him that the Coal Sack was God, and that he ought to tell someone.

Tradition had it that the Face of God could be seen from New Caledonia; and Littlemead had a powerful voice. Despite the opposition of the Imperial Orthodox Church, despite the protests of the Viceroy and the scorn of the University staff, the Church of Him spread until it was a power on New Scotland.

It was never large, but its members were fanatics; and they had the miracle of the Mote, which no scientist could explain. By 2895 the Church of Him was a power among New Scot farmers, but not in the cities. Still, half the population worked in the fields. The converter kitchens had all broken down.

By 2900 New Scotland had two working interplanetary spacecraft, one of which could not land. Its Langston Field had died. The term was appropriate. When a piece of Empire technology stopped working, it was dead. It could not be repaired. New Scotland was becoming primitive.

For forty years the Mote had grown. Children refused to believe that it had once been called the Mote. Adults knew it was true, but couldn't remember why. They called the twin stars Murcheson's Eye, and believed that the red supergiant had no special name.

The records might have showed differently, but the University records were suspect. The Library had been scrambled by electromagnetic pulses during the years of siege. It had large areas of amnesia.

In 2902 the Mote went out.

Its green light dimmed to nothing over a period of several hours; but that happened on the other side of the world. When the Coal Sack rose above University City that night, it rose as a blinded man.

All but a few remnants of the Church of Him died that year. With the aid of a handful of sleeping pills Howard Grote Littlemead hastened to meet his God... possibly to demand an explanation.

Astronomy also died. There were few enough astronomers and fewer tools; and when nobody could explain the vanishing of the Mote... and when telescopes turned on the Mote's remnant showed only a yellow dwarf star, with nothing remarkable about it at all....

People stopped considering the stars. They had a world to save.

The Mote was a G2 yellow dwarf, thirty-five light-years distant: a white point at the edge of Murcheson's Eye. So it was for more than a century, while the Second Empire rose from Sparta and came again to New Caledonia.

Then astronomers read old and incomplete records, and resumed their study of the red supergiant known as Murcheson's Eye; but they hardly noticed the Mote.

And the Mote did nothing unusual for one hundred and fifteen years.

---

Thirty-five light-years away, the aliens of Mote Prime had launched a light-sail spacecraft, using batteries of laser cannon powerful enough to outshine a neighboring red supergiant.

As for why they did it that way, and why it looked like that, and what the bejeesus is going on... explanations follow.

Most hard science-fiction writers follow standard rules for building worlds. We have formulae and tables for getting the orbits right, selecting suns of proper brightness, determining temperatures and climates, building a plausible ecology. Building worlds requires imagination, but a lot of the work is mechanical. Once the mechanical work is done the world may suggest a story, or it may even design its own inhabitants. Larry Niven's "known space" stories include worlds which have strongly affected their colonists.

Or the exceptions to the rules may form stories. Why does Mote Prime, a nominally Earthlike world, remind so many people of the planet Mars? What strangeness in its evolution made the atmosphere so helium-rich? This goes beyond mechanics.

In *The Mote in God's Eye* (Simon and Schuster, 1974) we built not only worlds, but cultures.

From the start *Mote* was to be a novel of first contact. After our initial story conference we had larger ambitions: *Mote* would be, if we could write it, the *epitome* of first contact novels. We intended to explore every important problem arising from first contact with aliens—and to look at those problems from both human and alien viewpoints.

That meant creating cultures in far more detail than is needed for most novels. It's easy, when a novel is heavy with detail, for the details to get out of hand, creating glaring inconsistencies. (If civilization uses hydrogen fusion power at such a rate that world sea level has dropped by two feet, you will not have people sleeping in abandoned movie houses.) To avoid such inconsistencies we worked a great deal harder developing the basic technologies of both the Motie (alien) and the human civilizations.

In fact, when we finished the book we had nearly as much unpublished material as ended up in the book. There are many pages of data on Motie biology and evolutionary history; details on Empire science and technology; descriptions of space battles, how worlds are terraformed, how light-sails are constructed; and although these background details affected the novel and dictated what we would actually write, most of them never appear in the book.

We made several boundary decisions. One was to employ the Second Empire period of Pournelle's future history. That Empire existed as a series of sketches with a loose outline of its history, most of it previously published. *Mote* had to be consistent with the published material.

Another parameter was the physical description of the aliens. Incredibly, that's all we began with: a detailed description of what became the prototype Motie, the Engineer: an attempt to build a nonsymmetrical alien, left over from a Niven story that never quite jelled. The history, biology, evolution, sociology, and culture of the Moties were extrapolated from that being's shape during endless coffee-and-brandy sessions.

That was our second forced choice. The Moties lived within the heart of the Empire, but had never been discovered. A simple explanation might have been to make the aliens a young civilization just discovering space travel, but that assumption contradicted Motie history as extrapolated from their appearance. We found another explanation in the nature of the Alderson Drive.

# EMPIRE TECHNOLOGY

The most important technological features of the Empire were previously published in other stories: the Alderson Drive and Langston Field.

Both were invented to Jerry Pournelle's specifications by Dan Alderson, a resident genius at Caltech's Jet Propulsion Laboratory. It had always been obvious that the Drive and Field would affect the cultures that used them, but until we got to work on *Mote* it wasn't obvious just how profound the effects would be.

## *THE ALDERSON DRIVE*

Every SF writer eventually must face the problem of interstellar transportation. There are a number of approaches. One is to deny faster-than-light travel. This in practice forbids organized interstellar civilizations.

A second approach is to ignore General and Special Relativity. Readers usually won't accept this. It's a cop-out, and except in the kind of story that's more allegory than science fiction, it's not appropriate.

Another method is to retreat into doubletalk about hyperspace. Doubletalk drives are common enough. The problem is that when everything is permitted, nothing is forbidden. Good stories are made when there are difficulties to overcome, and if there are no limits to "hyperspace travel" there are no real limits to what the heroes and villains can do. In a single work the "difficulties" can be planned as the story goes along, and the drive then redesigned in rewrite; but we couldn't do that here.

Our method was to work out the Drive in detail and live with the resulting limitations. As it happens, the limits on the Drive influenced the final outcome of the story; but they were not invented for that purpose.

The Alderson Drive is consistent with everything now known about physics. It merely assumes that additional discoveries will be made in about thirty years, at Caltech (as a tip o' the hat to Dan Alderson). The key event is the detection of a "fifth force."

There are four known forces in modern physics: two sub-nuclear forces responsible respectively for alpha and beta decay; electromagnetism, which includes light; and gravity. The Alderson force, then, is the fifth, and it is generated by thermonuclear reactions.

The force has little effect in our universe; in fact, it is barely detectable. Simultaneously with the discovery of the fifth force, however, we postulate the discovery of a second universe in point-to-point congruence with our own. The "continuum universe" differs from the one we're used to in that there are no known quantum effects there.

Within that universe particles may travel as fast as they can be accelerated; and the fifth force exists to accelerate them.

There's a lot more, including a page or so of differential equations, but that's the general idea.

You can get from one universe to another. For every construct in our universe there can be created a "correspondence particle" in the continuum universe. In order for your construct to go into and emerge from the continuum universe without change you must have some complex machinery to hold everything together and prevent your ship and crew—from being disorganized into elementary particles.

Correspondence particles can be boosted to speeds faster than light: in fact, to speeds nearly infinite as we measure them. Of course they cannot emerge into our universe at such speeds: they have to lose their energy to emerge at all. More on that in a moment.

There are severe conditions to entering and leaving the continuum universe. To emerge from the continuum universe you must exit with precisely the same potential energy (measured in terms of the fifth force, not gravity) as you entered. You must also have zero kinetic energy relative to a complex set of coordinates that we won't discuss here.

The fifth force is created by thermonuclear reactions: generally, that is, in stars. You may travel by using it, but only along precisely defined lines of equipotential flux: tramways or tramlines.

Imagine the universe as a thin rubber sheet, very flat. Now drop heavy rocks of different weights onto it. The rocks will distort the sheet, making little cone-shaped (more or less) dimples. Now put two rocks reasonably close together: the dimples will intersect in a valley. The intersection will have a "pass," a region higher than the low points where the rocks (stars) lie, but lower than the general level of the rubber sheet.

The route from one star to another through that "pass" is the tramline. Possible tramlines lie between each two stars, but they don't always exist, because when you add third and fourth stars to the system they may interfere, so there is no unique gradient line. If this seems confusing, don't spend a lot of time worrying about it; we'll get to the effects of all this in a moment.

You may also imagine stars to be like hills; move another star close and the hills will intersect. Again, from summit to summit there will be one and only one line that preserves the maximum potential energy for that level. Release a marble on one hill and it will roll down, across the saddle, and up the side of the other. That too is a tramline effect. It's generally easier to think of the system as valleys rather than hills, because to travel from star to star you have to get over that "hump" between the two. The fifth force provides the energy for that.

You enter from the quantum universe. When you travel in the continuum universe you continually lose kinetic energy; it "leaks." This can be detected in our universe as photons. The effect can be important during a space battle. We cut such a space battle from *Mote*, but it still exists, and we may yet publish it as a novella.

To get from the quantum to the continuum universe you must supply power, and this is available only in quantum terms. When you do this you turn yourself into a correspondence particle; go across the tramline; and come out at the point on the other side where your potential energy is equal to what you entered with, plus zero kinetic energy (in terms of the fifth force and complex reference axes).

For those bored by the last few paragraphs, take heart: we'll leave the technical details and get on with what it all means.

Travel by Alderson Drive consists of getting to the proper Alderson Point and turning on the Drive. Energy is used. You vanish, to reappear in an immeasurably short time at the Alderson Point in another star system some several light-years away. If you haven't done everything right, or aren't at the Alderson Point, you turn on your Drive and a lot of energy vanishes. You don't move. (In fact you do move, but you instantaneously reappear in the spot where you started.)

That's all there is to the Drive, but it dictates the structure of an interstellar civilization.

To begin with, the Drive works only from point to point across interstellar distances. Once in a star system you must rely on reaction drives to get around. There's no magic way from, say, Saturn to Earth: you've got to slog across.

Thus space battles are possible, and you can't escape battle by vanishing into hyperspace, as you could in future history series such as Beam Piper's and Gordon Dickson's. To reach a given planet you must travel across its stellar system, and you must enter that system at one of the Alderson Points. There won't be more than five or six possible points of entry, and there may only be one.

Star systems and planets can be thought of as continents and islands, then, and Alderson Points as narrow sea gates such as Suez,

Gibraltar, Panama, Malay Straits, etc. To carry the analogy further, there's telegraph but no radio: the fastest message between star systems is one carried by a ship, but within star systems messages go much faster than the ships....

Hmm. This sounds a bit like the early days of steam. NOT sail; the ships require fuel and sophisticated repair facilities. They won't pull into some deserted star system and rebuild themselves unless they've carried the spare parts along. However, if you think of naval actions in the period between the Crimean War and World War One, you'll have a fair picture of conditions as implied by the Alderson Drive.

The Drive's limits mean that uninteresting stellar systems won't be explored. There are too many of them. They may be used as crossing-points if the stars are conveniently placed, but stars not along a travel route may never be visited.

Reaching the Mote, or leaving it, would be damned inconvenient. Its only tramline reaches to a star only a third of a light-year away—Murcheson's Eye, the red supergiant—and ends deep inside the red-hot outer envelope. The aliens' only access to the Empire is across thirty-five light-years of interstellar space—which no Empire ship would ever see. The gaps between the stars are as mysterious to the Empire as they are to you.

### *THE LANGSTON FIELD*

Our second key technological building block was the Langston Field, which absorbs and stores energy in proportion to the fourth power of incoming particle energy: that is, a slow-moving object can penetrate it, but the faster it's moving (or hotter it is) the more readily it is absorbed.

(In fact it's not a simple fourth-power equation; but surely you don't need third-order differential equations for amusement.)

The Field can be used for protection against lasers, thermonuclear weapons, and nearly anything else. It isn't a perfect defense, however. The natural shape of the Field is a solid. Thus it wants to collapse and vaporize everything inside it. It takes energy to maintain a hole inside the Field, and more energy to open a control in it so that you can cause it selectively to radiate away stored energy. You don't get something for nothing.

This means that if a Field is overloaded, the ship inside vanishes into vapor. In addition, *parts* of the Field can be momentarily overloaded: a sufficiently high energy impacting a small enough area will cause a temporary Field collapse, and a burst of energy penetrates to the inside. This can damage a ship without destroying it.

### *ASTROGRAPHY*

We've got to invent a term. What is a good word to mean the equivalent of "geography" as projected into interstellar space? True, planetologists have now adopted "geology" to mean geophysical sciences applied to any planet, not merely Earth; and one might reasonably expect "geography" to be applied to the study of physical features of other planets—but we're concerned here with the relationship of star systems to each other.

We suggest cosmography, but perhaps that's too broad? Should that term be used for relationships of *galaxies*, and mere star system patterns be studied as "astrography"? After all, "astrogator" is a widely-used term meaning "navigator" for interstellar travel.

Some of the astrography of *Mote* was given because it had been previously published. In particular, the New Caledonia system, and the red supergiant known as Murcheson's Eye, had already been worked out. There were also published references to the history of New Caledonia.

We needed a red supergiant in the Empire. There's one logical place for that, and previously published stories had placed one there:

Murcheson's Eye, behind the Coal Sack. It *has* to be behind the Coal Sack: if there were a supergiant that close anywhere else, we'd see it now.

Since we had to use Murcheson's Eye, we had to use New Caledonia. Not that this was any great imposition: New Scotland and New Ireland are interesting places, terraformed planets, with interesting features and interesting cultures.

There was one problem, though: New Scotland is inhabited by New Scots, a people who have preserved their sub-culture for a long time and defend it proudly. Thus, since much of the action takes place on New Scotland, some of the characters, including at least one major character, *had* to be New Scot. For structural reasons we had only two choices: the First Officer or the Chief Engineer.

We chose the Chief Engineer, largely because in the contemporary world it is a fact that a vastly disproportionate number of ship's engineers are Scots, and that seemed a reasonable thing to project into the future.

Alas, some critics have resented that, and a few have accused us of stealing Mr. Sinclair from *Star Trek.* We didn't. Mr. Sinclair is what he is for perfectly sound astrographical reasons.

The astrography eventually dictated the title of the book. Since most of the action takes place very near the Coal Sack, we needed to know how the Coal Sack would look close up from the back side. Eventually we put swirls of interplanetary dust in it, and evolving proto-stars, and all manner of marvels; but those came after we got *very* close. The first problem was the Coal Sack seen from ten parsecs.

Larry Niven hit on the happy image of a hooded man, with the supergiant where one eye might be. The supergiant has a small companion, a yellow dwarf not very different from our Sun. If the supergiant is an eye—Murcheson's Eye—then the dwarf is, of course, a mote in that eye.

But if the Hooded Man is seen by backward and superstitious peoples as the Face of God... then the name for the Mote becomes inevitable… and once suggested, "The Mote in God's Eye" is a near irresistible title. (Although in fact Larry Niven did resist it, and wanted "The Mote in Murcheson's Eye" up to the moment when the publisher argued strongly for the present title….)

### *THE SHIPS*

Long ago we acquired a commercial model called "The Explorer Ship *Leif Ericson*," a plastic spaceship of intriguing design. It is shaped something like a flattened pint whiskey bottle with a long neck. The *Leif Ericson*, alas, was killed by general lack of interest in spacecraft by model buyers; a ghost of it is still marketed in hideous glow-in-the-dark color as some kind of flying saucer.

It's often easier to take a detailed construct and work within its limits than it is to have too much flexibility. For fun we tried to make the *Leif Ericson* work as a model for an Empire naval vessel. The exercise proved instructive.

First, the model is of a *big* ship, poorly designed in shape ever to be carried aboard another vessel. Second, it had fins. Fins are only useful for atmosphere flight: what purpose would be served in having atmosphere capabilities on a large ship?

This dictated the class of ship. It must be a cruiser or battlecruiser. Battleships and dreadnoughts wouldn't ever land, and would be cylindrical or spherical to reduce surface area. Our ship was too large to be a destroyer (an expendable ship almost never employed on missions except as part of a flotilla). Cruisers and battlecruisers can be sent on independent missions.

*MacArthur*, a General Class Battlecruiser, began to emerge. She can enter atmosphere, but rarely does so, except when long independent assignments force her to seek fuel on her own. She can do this in either of two ways: go to a supply source, or fly into the hydrogen-rich atmosphere of a gas giant and scoop. There were scoops on the model, as it happens.

She has a large pair of doors in her hull, and a spacious compartment inside: obviously a hangar deck for carrying auxiliary craft. Hangar deck is also the only large compartment in her, and therefore would be the normal place of assembly for the crew when she isn't under battle conditions.

The tower on the model looked useless, and was almost ignored, until it occurred to us that on long missions not under, acceleration it would be useful to have a high-gravity area. The ship is a bit thin to have much gravity in the "neck" without spinning her far more rapidly than you'd like; but with the tower, the forward area gets normal gravity without excessive spin rates.

And on, and so forth. In the novel, *Lenin* was designed from scratch; and of course we did have to make some modifications in *Leif Ericson* before she could become *INSS MacArthur*, but it's surprising just how much detail you can work up through having to live with the limits of a model....

## SOCIOLOGY

The Alderson Drive and the Langston Field determine what kinds of interstellar organizations will be possible. There will be alternatives, but they have to fit into the limits these technologies impose.

In *Mote* we chose Imperial Aristocracy as the main form of human government. We've been praised for this: Dick Brass in a *New York Post* review concludes that we couldn't have chosen anything else, and other critics have applauded, us for showing what such a society might be like.

Fortunately there are no Sacred Cows in science fiction. Maybe we should have stuck to incest? Because other critics have been horrified! Do we, they ask, really *believe* in imperial government? and *monarchy*?

That depends on what they mean by "believe in." Do we think it's desirable? We don't have to say. Inevitable? Of course not. Do we think it's *possible*? Damn straight.

The political science in *Mote* is taken from C. Northcote Parkinson's *Evolution of Political Thought.* Parkinson himself echoes Aristotle.

It is fashionable to view history as a linear progression: things get better, never worse, and of course we'll never go back to the bad old days of (for instance) personal government. Oddly enough, even critics who have complained about the aristocratic pyramid in *Mote*—and thus rejected our Empire as absurd—have been heard to complain about "Imperial presidency" in the U.S.A. How many readers would bet long odds against John-John Kennedy becoming president within our lifetimes?

*Any* pretended "science" of history is the bunk. That's the problem with Marxism. Yet Marx wrote a reasonable view of history up to his time, and some of his principles may be valid.

Military history is another valid way to view the last several thousand years—but no one in his right mind would pretend that a history of battles and strategies is the whole of the human story. You may write history in terms of medical science, in terms of rats, lice, and plagues, in terms of agricultural development, in terms of strong leadership personalities, and each view will hold some truth.

There are many ways to view history, and Aristotle's cycles as brought up to date by Parkinson make one of the better ones. For those who don't accept that proposition, we urge you at least to read Parkinson before making up your minds and closing the door.

The human society in *Mote* is colored by technology and historical evolution. In *Mote's* future history the United States and the Soviet Union form an alliance and together dominate the world during the last decades of the 20th century. The alliance doesn't end their rivalry, and doesn't make the rulers or people of either nation love their partners.

The CoDominium Alliance needs a military force. Military people need something or someone they can give their loyalty; few men ever risked their lives for a "standard of living" and there's little that's more stupid than dying for one's standard of living—unless it's dying for someone else's standard of living.

Do the attitudes of contemporary police and soldiers lead us to suppose that "democracy" or "the people" inspire loyalty? The proposition is at least open to question. In the future that leads to *Mote*, a Russian admiral named Lermontov becomes leader of CoDominium forces, and although he is not himself interested in founding a dynasty, he transfers the loyalty of the Fleet to leaders who are.

He brings with him the military people at a time of great crisis. Crises have often produced strong loyalties to single leaders:

Churchill, Roosevelt, George Washington, John F. Kennedy during the Cuban Crisis, etc. (A year after Kennedy's death Senator Pastore could address a national convention and get standing ovations with the words "There stood John Kennedy, TEN FEET TALL!!!")

Thus develops the Empire.

Look at another trend: personal dictatorship. There were as many people ruled by tyrants as by "democracy" in 1975, and even in the democracies charges of tyranny are not lacking. Dictatorships may not be the wave of the future—but is it unreasonable to suppose they might be?

Dictatorship is often tried in times of severe crisis: energy crisis, population crisis, pollution crisis, agricultural crisis—surely we do not lack for crises? The trouble with dictatorship is that it generates a succession crisis when the old man bows out. Portugal, for one, has gone through such a period. Chile, Uganda, Brazil, name your own examples: anyone want to bet that some of these won't turn to a new Caudillo with relief?

How to avoid succession crisis? One traditional method is to turn Bonapartist: give the job to a relative or descendant of the dictator. He may not do the job very well, but after enough crises people are often uninterested in whether the land is governed well. They just want things *settled* so they can get on with everyday life.

Suppose the dictator's son does govern well? A new dynasty is founded, and the trappings of legitimacy are thrust onto the new royal family. To be sure, the title of "King" may be abandoned. Napoleon chose to be "Emperor of the French," Cromwell chose "Lord Protector," and we suppose the U.S. will be ruled by presidents for a long time—but the nature of the presidency, and the way one gets the office, may change.

See, for example, Niven's use of "Secretary-General" in the tales of Svetz the time-traveller.

We had a choice in *Mote*: to keep the titles as well as the structure of aristocratic empire, or abandon the titles and retain the structure only. We could have abolished "Emperor" in favor of "President," or "Chairperson," or "Leader." or "Admiral," or "Posnitch." The latter, by the way, is the name of a particularly important president honored for all time by having his name adopted as the title for Leader....

We might have employed titles other than Duke (originally meant "leader" anyway) and Count (Companion to the king) and Marquis (Count of the frontier marches). Perhaps we should have. But any titles used would have been *translations* of whatever was current in the time of the novel, and the traditional titles had the effect of letting the reader know quickly the approximate status and some of the duties of the characters.

There are hints all through *Mote* that the structure of government is not a mere carbon copy of the British Empire or Rome or England in the time of William III. On the other hand there are similarities, which are forced onto the Empire by the technology we assumed.

Imperial government is not inevitable. It is possible.

The alternate proposition is that today we are so advanced that we will never go back to the bad old days. Yet we can show you essays "proving" exactly that proposition—and written thousands of years ago. There's a flurry of them every few centuries.

We aren't the first people to think we've "gone beyond" personal government, personal loyalties, and a state religion. Maybe we won't be the last.

Anyway, *Mote* is supposed to be entertainment, not an essay on the influence of science on social organization. (You're getting *that* here.)

The Empire is what it is largely because of the Alderson Drive and Langston Field. Without the Drive an empire could not form. Certainly an interstellar empire would look very different if it had to depend on lightspeed messages to send directives and receive reports. Punitive expeditions would be nearly impossible, hideously expensive, and probably futile: you'd be punishing the grandchildren of a generation that seceded from the Empire, or even a planet that put down the traitors after the message went out.

Even a rescue expedition might never reach a colony in trouble. A coalition of bureaucrats could always collect the funds for such an expedition, sign papers certifying that the ships are on the way, and pocket the money… in sixty years someone might realize what had happened, or not.

The Langston Field is crucial to the Empire, too. The Navy can survive partial destruction and keep fighting. Ships carry black boxes—plug-in sets of spare parts—and large crews who have little to do unless half of them get killed. That's much like the navies of fifty years ago.

A merchant ship might have a crew of forty. A warship of similar size carries a crew ten times as large. Most have little to do for most of the life of the ship. It's only in battles that the large number of self-programming computers become important. *Then* the outcome of the battle may depend on having the largest and best-trained crew—and there aren't many prizes for second place in battle.

Big crews with little to do demand an organization geared to that kind of activity. Navies have been doing that for a long time, and have evolved a structure that they tenaciously hold onto.

Without the Field as defense against lasers and nuclear weapons, battles would become no more than offensive contests. They'd last microseconds, not hours. Ships would be destroyed or not, but hardly ever wounded. Crews would tend to be small, ships would be different, including something like the present-day aircraft carriers. Thus technology dictates Naval organization.

It dictates politics, too. If you can't get the populace, or a large part of it, under a city-sized Field, then any given planet lies naked to space.

If the Drive allowed ships to sneak up on planets, materializing without warning out of hyperspace, there could be no Empire even with the Field. There'd be no Empire because belonging to an Empire wouldn't protect you. Instead there might be populations of planet-bound serfs ruled at random by successive hordes of space pirates. Upward mobility in society would consist of getting your own ship and turning pirate.

Given Drive and Field, though, Empires are possible. What's more likely? A representative confederacy? It would hardly inspire the loyalty of the military forces, whatever else it might do. (In the War Between the States, the Confederacy's main problem was that, the troops were loyal to their own State, not the central government.)

Each stellar system independent? That's reasonable, but is it stable? Surely there might be pressures toward unification of at least parts of interstellar space.

How has unification been achieved in the past? Nearly always by conquest or colonization or both. How have they been held together? Nearly always by loyalty to a leader, an emperor, or a dynasty, generally buttressed by the trappings of religion and piety. Even Freethinkers of the last century weren't ashamed to profess loyalty to the Widow of Windsor....

Government over large areas needs emotional ties. It also needs *stability*. Government by 50%-plus-one hasn't enjoyed particularly stable politics—and it lasts only so long as the 50%-minus-one minority is willing to submit. Is heredity a rational way to choose leaders? It has this in its favor: the leader is known from an early age to be destined to rule, and can be educated to the job. Is that preferable to education based on how to *get* the job? Are elected officials better at governing, or at winning elections?

Well, at least the counter-case can be made. That's all we intended to do. We chose a stage of Empire in which the aristocracy was young and growing and dynamic, rather than static and decadent; when the aristocrats are more concerned with duty than with privilege; and we made no hint that we thought that stage would last forever.

## RANDOM DETAILS

Robert Heinlein once wrote that the best way to give the flavor of the future is to drop in, without warning, some strange detail. He gives as an example, "The door dilated."

We have a number of such details in *Mote*. We won't spoil the book by dragging them all out in a row. One of the most obvious we use is the personal computer, which not only does computations, but also puts the owner in contact with any nearby data bank; in effect it will give the answer to any question whose answer is known and that you think to ask.

Thus no idiot block gimmicks in *Mote*. Our characters may fail to guess something, or not put information together in the right way, but they won't *forget* anything important. The closest that comes to happening is when Sally Fowler can't quite remember where she filed the tape of a conversation, and she doesn't take long to find it then.

On the other hand, people can be swamped with too much information, and that does happen.

There were many other details, all needed to keep the story moving. A rational kind of space suit, certainly different from the clumsy things used now. Personal weapons. The crystal used in a banquet aboard *MacArthur*: crystal strong as steel, cut from the windshield of a wrecked First Empire re-entry vehicle, indicating the higher technology lost in that particular war. Clothing and fashion; the status of women; myriads of details of everyday life.

Not that *all* of these differ from the present. Some of the things we kept the same probably will change in a thousand years. Others... well, the customs associated with wines and hard liquors are old and stable. If we'd changed everything, and made an attempt to portray every detail of our thousand-year-advanced future, the story would have gotten bogged down in details.

*Mote* is probably the only novel ever to have a planet's orbit changed to save a line.

New Chicago, as it appeared in the opening scenes of the first draft of *Mote*, was a cold place, orbiting far from its star. It was never a very important point, and Larry Niven didn't even notice it.

Thus when he introduced Lady Sandra Liddell Leonovna Bright Fowler, he used as viewpoint character a Marine guard sweating in hot sunlight. The Marine thinks, "She doesn't sweat. She was carved from ice by the finest sculptor that ever lived."

Now that's a good line. Unfortunately it implies a hot planet. If the line must be kept, the planet must be moved.

So Jerry Pournelle moved it. New Chicago became a world much closer to a somewhat cooler sun. Its year changed, its climate changed, its whole history had to be changed....

Worth it, though. Sometimes it's easier to build new worlds than to think up good lines....

# THAT BUCK ROGERS STUFF

The young lady was very serious and although I might have wished that she were an ogre, with raucous voice, and nose meeting chin in front of her lips, she was actually very professional in appearance; highly attractive, and—according to most objective standards—intelligent.

My wife and I had come to a typical Los Angeles show-business party. The young lady had been waiting for me. Before I could get properly into the room she advanced menacingly.

"You write science fiction," she accused. "Escapism. What good does it do to get people dreaming about that Buck Rogers stuff?" (I swear it, she used that phrase, the same one that countless teachers used in the days of my youth when they caught me reading *Astounding Science Fiction.*)

Naturally, she had A Cause. "We spent billions for what? For some pieces of rock and pretty pictures on television! And there are millions out of jobs, we need better schools, and–"

Some of you have probably had similar experiences and can finish off the speech for yourself. It's not the only time I've been put to The Question: "Why throw money away on space when there's so much that needs doing here on Earth?" All right, let's talk about space and see just how far we can get.

First, a couple of commercials. For a really beautiful job of discussing what we've *already* got out of space, send to NASA, Washington DC 20546 and request a copy of *Spinoff 1976*. My copy has no price on it; I got it as a gift from the National Space Institute (1911 Fort Myer Drive, Suite 408, Arlington VA 22209, dues $15 annually, $9 for students, and if you haven't joined yet, DO IT!). I expect NASA has some nominal charge for *Spinoff 1976*, but you could probably get one free through your Congress-critter.

*Spinoff* was written by Neil Rusczic of NSI. He's also the author of an excellent book called *Where the Winds Sleep*, something else I recommend. Between *Spinoff*, and Rusczic's book you can find plenty answers to the silly question about why spend money on space.

In fact, the problem is knowing where to begin. Weather predictions? Remember when the weatherman was a joke? True, the Weather Bureau makes some mistakes even yet; but not very many, and almost never when it comes to hurricanes. You can show that the space program has pretty well paid for itself just in better weather forecasting alone.

Those concerned about pollution will be pleased to hear that Earthwatch satellites finally give us a chance to see the real effects of pollution. Mining prospecting has been revolutionized by satellite photography. The international Food and Agricultural Organization in Rome can, from satellite data, get a good forecast of famine areas and global food production.

That's all satellite stuff. Industry benefits are nearly incalculable, and I don't mean frivolities like Teflon frying pans. Stuff like test procedures and quality control: the inspection methods developed for man-rating spacecraft and boosters are now routinely used in building better plows, tractors, automobiles, skis, hiking boots and packframes, electronic equipment, and darned near anything else you can think of.

In my early days in the space program one of the hardest jobs we had was monitoring physiological conditions in a stress environment. Just getting an ordinary electrocardiograph (EKG) through a pressure wall required great ingenuity. We invented a number of such devices; we had to. My own inventions are long since obsolete—but the space medicine technology that grew out of our early efforts is routinely used in hospitals and clinics all over the world. Mass spectrometers to analyze exhaled breath; microminiature EKG systems worn by hospital patients and displaying abnormalities to the duty nurse; blood analysis equipment; even heart condition diagnosis from moving vehicles; all routine, and all developed as part of the NASA package.

Your tires last longer, you can buy large fiberglass structures, firemen can keep your house from burning, your electrical system is simpler, crash helmets work better (remind me sometime to tell you about the purchase order for "nine freshly-killed human male corpses, ages 21 to 40 at time of death, must not have any abnormalities of brain or upper spine; expendable research item; no salvage value." The Purchasing Officer's reaction to that was, uh, interesting), driver-training simulators work, paints last longer, and golf clubs do a better job of driving the ball.

"Whoa. That's all technology, and technology is evil. It causes pollution, and kills people in wars, and–"

And at that point my usual reaction is a loud "Aaargh!" and a burning desire to find a drink. Quickly. Especially when it was said by a young person wearing a thin wristwatch and polyester imitations of honest blue denim, driving a Mercedes, and feeling committed because she hasn't eaten table grapes for *weeks*. I should control that reaction, of course; but if I were able to do that I'd probably still be in aerospace management instead of living the unnatural life of a writer.

Still, such people ought to be answered. Our whole future may depend on it. Let's try.

California's Governor Jerry Brown has built himself quite a reputation by pushing "Alternate Technology" and the philosophy that goes with it. "Make do. Expect less. Conserve. Smaller is better. Recycle. Be satisfied with what you have. There's Only One Earth."

Now there are some attractive points about all that. Moreover, the vision of a stable, low-to-zero-growth economy, concentrating on adventures of the mind, with a lot of "cottage industry" can be a noble one. It's probably possible, too—for us, and for a while.

It is not a philosophy likely to appeal to the poor of this world. Like it or not, a conservation-oriented low-growth world economy dooms most of the world's people to wretched poverty. But what has that to do with *us*? Can we not, ourselves, change our ways and let others go theirs?

Probably not. Like it or not, we've got most of the technology—and we don't have enough to develop the Earth to a point of satiation. If all the world gets rich through the same wasteful processes we employed, we're probably in big trouble. Worse, what of our grandchildren? The Earth's resources will not last forever; and what then?

I've argued here before that this generation is crucial: we have the resources to get mankind off this planet. If we don't do it, we may soon be facing a world of 15 billion people and more, a world in which it's all we can do to stay alive; a world without the investment resources to go into space and get rich. Usually I think it won't come to that; it's only in odd moments—such as when faced with The Question—that I get depressed.

I don't think it will come to that, because the vision of the future is so clear to me.

We need realize only one thing: we do not inhabit "Only One Earth."

Mankind doesn't live on Earth. We live in a solar system of nine planets, 34 moons, and over half a million asteroids. That system circles a rather small and unimportant star that is part of a galaxy containing tens of billions of stars. Only One Earth, indeed! There are millions of Earths out there, and if we use up this one, we'll just have to go find another, that's all.

We needn't use up this one. In "Survival With Style" I went through the numbers: how we can, with present-day technology, deliver here to Earth as much metal for each person in the world as the U.S. disposed of per capita in the 60's. We can do that without polluting our planet at all, and we can keep it up for tens of thousands of years. The metal is out there in the asteroid belt. For starters we don't even have to look very hard; most of the asteroids were once spherical, large enough to have metallic cores, and now the worthless gubbage topside has been knocked away, exposing all that lovely iron and lead and tin and such we'll need to give the wretched of the Earth *real* freedom.

Why not? The refinery power's there; the Sun gives it off for free. We have a propulsion system to get us to the asteroids; Project NERVA was cancelled, but the research was done, and it wouldn't be that hard to start up again. Nuclear-powered rockets would be rather simple to build, if we wanted them.

But first we'll need a Moonbase. We can get that the hard way, carrying stuff up bit by bit from the top of disintegrating totem poles, but there are easier ways.

We could do it in one whack. Project ORION was also cancelled, but we could build old Bang-Bang in a very few years if we wanted to. ORION used the simplest and most efficient method of nuclear propulsion of all: take a BIG plate, quite thick and hard; attach by shock-absorbers a large space-going capsule to it; put underneath one each atomic bomb; and fire away.

Believe me, your ship will move. When you've used up the momentum imparted by the first bomb, fling another down underneath. Repeat as required. For the expenditure of a small part of the world's nuclear weapon stockpile you have put several *million* pounds into orbit, or on the Lunar surface.

But that will cause fallout.

Yes; some. Not very much, compared to what we have already added to background radiation, but perhaps enough that we don't want to use ORION—although, he said happily, ORION is one reason why I think we'll eventually do what has to be done, even if this generation fails in its duties to the future. ORION is cheap and the bombs won't go away; if we're still alive in that grim world of 15–20 billion and no space program, *somebody's* going to revive Bang-Bang and get out there.

ORION gets a few big payloads to orbit or the Moon. A more systematic way would be to build a big laser launching system and make it accessible to anyone with a payload to put into orbit. Freeman Dyson calls laser launch systems "space highways." The government builds the launch system, and can use it for its own purposes; but it also gives private citizens, consortiums, firms, a means of reaching orbit.

Dyson envisions a time when individual families can buy a space capsule, and once Out There, can do as they like. Settle on the Moon, stay in orbit, go find an asteroid; whatever. It will be a while before we can build cheap self-contained space capsules operable by the likes of you and me; but it may not be anywhere near as long as you think.

The problem is the engines, of course; there's nothing else in the space home economy that couldn't, at least in theory, be built for about the cost of a family home car, and recreational vehicle. But then most land-based prefabricated homes don't have their own motive power either; they have to hire a truck for towing.

It could make quite a picture: a train of space capsules departing Earth orbit for Ceres and points outward, towed by a ship something like the one described in my story "Tinker." Not quite Ward Bond in *Wagon Train*, but it still could make a good TV series. The capsules don't have to be totally self-sufficient, of course. It's easy enough to imagine way stations along the route, the space equivalent of filling stations in various orbits.

Dyson is fond of saying that the U.S. wasn't settled by a big government settlement program, but by individuals and families who often had little more than courage and determination when they started. Perhaps that dream of the ultimate in freedom is too visionary; but if so, it isn't because the technology won't exist.

However we build our Moonbase, it's a very short step from there to asteroid mines. Obviously the Moon is in Earth orbit; with the shallow Lunar gravity well it's no trick at all to get away from the Moon, and Earth orbit is halfway to anywhere in the solar system. We don't know what minerals will be available on the Moon. Probably it will take a while before it gets too expensive to dig them up, but as soon as it does, the Lunatics themselves will want to go mine the asteroids.

There's probably more water ice in the Belt than there is on Luna, so for starters there will be water prospectors moving about among the asteroids. The same technology that sends water to Luna will send metals to Earth orbit. I've already described one ship that can do the job. There are others. The boron fusion-fission process is a good example.

Take boron-11 ($^{11}B_5$). Bombard with protons. The result is a complex reaction that ends with helium and no nuclear particles. It could be a direct spacedrive. For those interested, the basic equation is

$$^{11}B_5 + p = 3(^4He_2) + 16\ \text{MeV} \qquad \text{(Eq. 8)}$$

and 16 million electron volts gives pretty energetic helium. The exhaust velocity is better than 10,000 kilometers/second, giving a theoretical specific impulse ($I_{sp}$ of something over a million. For comparison the $I_{sp}$ of our best chemical rockets is about 400, and NERVA manages something like 1200. The boron drive needn't be used very efficiently to send ships all over the solar system.

Meanwhile, NERVA or a fission-ion drive will do the job. In fact, it's as simple to get refined metals from the asteroid belt to near-Earth orbit as it is to bring them down from the Lunar surface. It takes longer, but who cares? If I can promise GM steel at less than they're now paying, they'll be glad to sign a "futures" contract, payment on delivery.

It's going to be colorful out in the Belt, with huge mirrors boiling out chunks from mile-round rocks, big refinery ships moving from rock to rock; mining towns, boom towns, and probably traveling entertainment vessels. Perhaps a few scenes from the Wild West? "Claim jumpers! Grab your rifles–"

Thus from the first Moonbase we'll move rapidly, first to establish other Moon colonies (the Moon's a *big* place) and out to the asteroid belt. After that we'll have fundamental decisions to make.

We can either build O'Neill colonies or stay with planets and moons. I suspect we'll do both. While one group starts constructing flying city-states at the Earth-Moon Trojan points, another will decide to make do with Mars.

Mars and Venus aren't terribly comfortable places; in fact, you probably won't want to land on Venus at all until it has been terraformed. Between Mars and Venus, Venus is the easiest to make into a shirt-sleeves inhabitable world. It requires only biological packages and some fertilizers and nutrients, and can be done from Moonbase, or in a pinch, from Earth itself.[4] Although Venus may be the simpler job, Mars is likely to come first, simply because you can live there before terraforming, and there will be people establishing dome colonies on the Red Planet.

[4] See "The Big Rain," Galaxy, 1975 September

I wrote a story (*Birth of Fire*) describing one Mars-terraforming project: melt the polar caps and activate a number of Martian volcanoes to get an atmosphere built up. Isaac Asimov described the final step many years ago: get your ice from Out There, at Jupiter or Saturn, and fling it downhill to Mars. Freeman Dyson points out that there's enough ice on Enceladus (a Saturnian moon) to keep the Martian climate warm enough for 10,000 years. The deserts of Mars can become gardens in less than a century.

Dyson's scheme didn't even involve human activity on Enceladus; robots and modern computers could probably accomplish the job. They've only to construct some big catapults on the surface of Enceladus, and build some solar sails. Dyson suggests robots because the project as described would take a long time, and human supervisors might not care for the work; but I suspect we could get plenty of volunteers if we needed them. Why not? No one could complain that the work was trivial, and you couldn't ask for an apartment with a better view than Saturn's Rings!

Moonbases. Lunar cities. Mining communities in the asteroid belt. Domed colonies on Mars, with prospects for terraforming the planet and turning it into a paradise. An advanced engineering project headquarters on Enceladus. Pollution controlled on Earth, because most polluting activities would go on in space. Near-Earth space factories. Several to hundreds of city-states at the Trojan points of the Earth-Moon system. A space population of millions, with manned and unmanned ships stitching all the space habitats together. This is not a dream world; this is a world we could make in a hundred years!

In 1872 a number of Kiowa and Comanche chiefs were taken to Washington by Quakers in an attempt to show the Indians just what they were facing. When they returned to talk about the huge cities, and "a stone tipi so large that all the Kiowa could sit under it," they were not believed. One suspects that if the Quaker schoolmasters had been magically transported to the Washington of 1976 and then returned to their own time, they would not be believed either. A nation of over 200 million people? Millions of tons of concrete poured into gigantic highways? Aircraft larger than the biggest sailing ships? City streets brightly lit at night? Millions of tons of steel, farmlands from Kansas to California....

Building a space civilization in the next hundred years will be simpler than getting where we are from 1876. We already know how to do it. We probably don't know how we *will* do it; certainly the very act of space exploration will generate new ideas and techniques as alien to us as nuclear energy would have been to Lord Rutherford or Benjamin Franklin; but we already know how we *could* do it. No basic new discoveries necessary.

In the 1940's I did a class report on space travel. I drew heavily from *Astounding*, from Heinlein's Future History, from Willy Ley's books on rockets and space travel. My teachers were tolerant. They let me do it. They didn't believe in suppressing their pupils. Afterwards, though, the physics teacher called me in for a conference: I should learn some good basic science, and get my head out of the clouds. That Buck Rogers stuff was fine for amusement—he read it himself—but in the real world....

In the real world I got a letter from that teacher, who had the honesty to send a note in August, 1969, apologizing to me and expressing gratitude that he'd not been able to discourage me from those crazy dreams. I wish he were alive so I could find out his reaction to *this* article.

It's not crazy dreams. It's not even Far Out. It's only basic engineering, and some economics, and a bit of hope. I may even have been too conservative. It probably won't take a hundred years.

Given the basic space civilization I've described, we'll have accomplished one goal: no single accident, no war, no one insane action will finish us off. We won't *have* to have outgrown our damn foolishness to insure survival of the race. Perhaps we'll all be adults, mature, satisfied with what we have, long past wars and conflicts and the like; but I doubt it. At least, though, there will be no way to exterminate mankind, even if we manage to make the Earth uninhabitable; and it's unlikely that any group, nation, or ideology can enslave everyone. That's worth something.

One suspects, too, that there will be an *enormous* diversity of cultures. Travel times between various city-states—asteroid, Martian, Lunar, O'Neill colony, Saturnian forward base, Jovian Trojan point—will be weeks to months to years with currently foreseeable technology. That's likely to change, but by the time the faster travel systems are in widespread use the cultural diversities will be established. Meanwhile, communication among all the various parts of the solar system will be simple and relatively cheap, so that there will have been that unifying influence; cultures will become different because people want to be different, not because they don't know any better.

Okay. In 100 years we'll have built a space civilization. We'll no longer have really grinding poverty, although there will undoubtedly be people who consider themselves poor, just as we have today people who while living better than the aristocrats of 1776 think themselves in terrible straits. We'll have insured against any man-made disaster wiping out the race. So what's next, besides more of the same?

Why, we haven't even got started yet! "Be fruitful and multiply, and fill the face of the Earth," said the command; soon that will have been done; and some day we'll even run up against a filled solar system.

The first step is obvious. We can begin taking some of the more useless planets apart. They've got all that lovely mass, and it's concentrated so that we can't use it; better to make proper use of, say, Jupiter, and Mercury, and someday perhaps even Mars and Venus despite our having terraformed them.

At a thousand tons of mass per person, Mercury, taken apart, could provide living space for $3 \times 10^{20}$ people—that's 300 billion

billion, rather a large population. People in the U.S. at present dispose of about $10^{18}$ ergs per capita each year; small potatoes for a space civilization. Let's figure that our space people will need a million times that much, $10^{24}$ ergs each per year, or a total of $3 \times 10^{44}$ ergs for the people living on the skeleton of Mercury.

It's too much. The Sun only puts out $2 \times 10^{39}$ ergs each year, and we can't catch all that. It seems we'll run out of energy before we run out of mass, and that mass is too handy to use up freely as energy. Back to energy conservation! To support a really large population, though, we'll have to destroy some matter. Obviously that can't go on forever: so, while we're destroying matter, we may as well go elsewhere.

Meanwhile, though, the stay-at-homes will busily take planets apart for their mass, so filling space with flying cities that they'll soon catch great quantities of solar energy. You can just hear the asteroid civilizations (what's left of them) complaining about those closer in taking up all the light. Perhaps the Rockrats will be the first to say the Hell with it and leave, looking for a place to live where there's *elbow room.* Just too crowded in the solar system. "Not like when I was a kid, Martha. Not room to swing a cat nowadays."

They can take their whole civilization with them. The negotiations may take some time; the homebodies aren't going to want to let all that nice matter leave the system forever. Perhaps the Rockrats will promise to send back a nice fat planet from wherever they're going.

It will take a while to pay off the debt, but they can pay it back with very high interest.

The trip will take many years, but so what? The Rockrats have taken their civilization with them. They'll miss the Sun, and by the time they arrive they'll have used up most of their asteroid, but by then people will live long lifetimes—and they'll darned well know how to exploit the new stellar system. "We'll do it right, Martha! None of those upstart places like Freedonia!"

Of course they'll already know about the planets in their new system. There's no real limit to the size of telescope you can build in space, and no problem about seeing; and with the lengthy baseline of the orbit of Ceres, or Jupiter's Trojans, or a Saturnian moon, astronomers will long since have discovered all the planets of all the nearby stars. There will probably have been probes sending back high-resolution pictures and making certain our colonists aren't heading for an already-occupied system.

And so it goes; across the galaxy, as mankind fills system after system, and somebody begins to feel crowded. You'll note I haven't even postulated faster-than-light travel; I have given us matter annihilation, although that's not strictly necessary.

And beyond that? When we've tapped all the resources of easily available planets, and are still running out of metals and just plain mass? Well, there are stars–

Take an old star. A red giant, perhaps. Useless. No planets left—all consumed in the nova explosion that formed an ordinary star into a red giant. The poor thing is doomed in a few million years anyway; why not hurry it along? When it blows up, it will give off all kinds of useful materials.

Of course the star is a long way from civilization. The minerals *could* be picked up after the explosion, but maybe there's a better way: bring your planet-sized spacecraft reasonably near the target star. Turn on the matter annihilators and focus the resulting energy into a rather powerful laser beam. Shine it properly on the star. That's what you're going to do to blow it up anyway, but if you're selective enough about it you can turn the star itself into a rocket. Heat up this side, let it spew out starstuff, and it will move. Granted that's a slow process, and perhaps there'll be no economic incentive; but stranger things have happened in history. After all, the expedition will save its parent civilization; and life aboard the control planet need not be any more dull than, say living in a colliery town; or going every day down to work at BBD&O....

But we needn't think about moving stars, or traveling to other stellar systems, any more than Columbus and the Vikings had Cape Canaveral in mind. For the moment we need only concentrate on the next hundred years. There's quite enough to do right here.

In fact, I can just hear it now: "What good does it do to get people dreaming about that Buck Rogers stuff? Why waste money on interstellar research when there's need for the money right here in the Trojan Points?"

Only One Earth indeed.

# BARDS OF THE SCIENCES

This article was originally published in the 1978 Science Fiction Writers Association Annual Magazine, with parodies of pulp SF magazine covers beautifully rendered in full-color by the late Robert Villani. In this volume, sadly, the cover illustrations can only be reproduced in black and white versions. A eulogy for Mr. Villani written by his daughter is included as an appendix.

---

*"Gather round," said the storyteller, "and I'll sing to you of humans who wrote fiction about science and themselves. And how what they wrote changed what they wrote about."*

Much of mankind—certainly anyone of a culture that takes an interest in the events in this yearbook—lives in a science fiction world. That statement is true on several levels. As the mass entertainment field enters the 1980s, Star Wars ranks as one of the all-time money-making motion pictures. Hundreds of science fiction titles are published each month. Science fiction magazines proliferate like desert flowers after rain. A whole generation is growing up watching *Star Trek* reruns on television. Not only is science fiction popular, but it also has become respectable, a development that some of its proponents find pleasant but quite surprising. Not very long ago science fiction was "Buck Rogers stuff"; now there are university courses in the genre. Yet there is a deeper sense in which people live in a science fiction world, for the world around them, the real world of everyday life, was science fiction not many years ago—and science fiction can claim to have had a major influence in creating it.

## THE ELUSIVE DEFINITION

There are currently 500 members of the Science Fiction Writers of America and many more practitioners in other countries. There are also several hundred academic teachers of SF. (Most science fiction authors prefer the abbreviation SF for science fiction, rather than the popular term sci-fi.) Although most have tried at one time or another to define science fiction, probably no two have ever agreed. Is there a difference, for instance, between "science fiction" and "speculative fiction"? Does science fiction include disaster stories set in the near future? Two examples of this type are J. G. Ballard's *The Crystal World*, in which all life on the Earth begins to crystallize, and *Lucifer's Hammer*, by Larry Niven and Jerry Pournelle, which describes the Earth's encounter with a large meteor. And what about fantasy, sword and sorcery, the ghost story, the fairy tale, or such social satire as Jonathan Swift's *Gulliver's Travels*? John W. Campbell, Jr., whose influence on modern science fiction was greater than that of any other editor, once defined science fiction as the real mainstream of literature, with everything else—including that which most people call mainstream—as a subcategory.

Defining science fiction is an impossible task, but certainly some agreement is needed in order to consider the impact of science fiction on science and society. Philosopher José Ortega y Gasset wrote that "to define is to exclude"; accordingly, this article will ignore fairy tales and fantasy and will be concerned primarily with what is usually called hard-science SF; i.e., stories that attempt, within limits, to be faithful to known laws of science. Nevertheless, other kinds of stories will inevitably enter into the discussion.

## A POOR PROPHET

Although often popularly supposed to predict the future, science fiction has seldom done that, nor do most science fiction writers claim any such ability. True, many SF editors of the so-called classic era from the mid-1920s to the early 1940s, particularly *Amazing Stories* magazine editor Hugo Gernsback, insisted on up-to-the-minute scientific accuracy and encouraged writers to project technological advances into the future. True, there have been numerous instances of correct "predictions," the most dramatic resulting in the 1944 visit by Manhattan Project security officers to *Astounding Science Fiction* magazine editor Campbell to discuss an alarmingly accurate fictional description of atomic energy. And true, the tradition of projecting technology into the future remains an important technique of modern science fiction. Yet, for all that, few SF authors would claim either the ability or the intent to predict the future.

Consider why any such claims would be untenable. The predictions of science fiction resemble those of the fortune-teller. A great many prophecies are made. Most are ambiguous, so that a large range of events can be counted "successful predictions"; these "successes" are loudly trumpeted while the many more numerous failures are conveniently forgotten. If science fiction is trying to predict, then the literature has a dismal record, not much better than the average soothsayer. There would be little point to examining its impact.

It can likewise be argued, on a case-by-case basis, that the best SF is not deliberately trying to predict. One example is Frank Herbert's *Dune*, a work remarkable for its realistic, detailed evocation of a desert planet, including its physiography, ecology, and human society. Nevertheless, it is this very complexity—this dependence on the interplay of so many coincident events—that makes it highly unlikely that the combination of environment and society in *Dune* will ever come to pass. It is difficult to suppose that Herbert could have crafted such a world without this understanding. Moreover, in some cases the book taxes known science; for instance, in its assumption that a breathable atmosphere could be created and maintained on a planet lacking water or some other source of oxygen. *The Mote in God's Eye*, by Niven and Pournelle, postulates an intelligent alien race that has altered its own evolution in a highly unexpected way, but its authors would be much surprised if future astronauts found any such creatures. Yet these are examples of two hard-science SF novels. A great deal of SF is not hard science; many authors have never attempted to make their works consistent with known science and technology.

Even hard-science SF authors have written nonscientific SF stories. As an example; Poul Anderson, justly known for stories incorporating the very latest in scientific developments, also wrote *Three Hearts and Three Lions*, in which a 20th-century engineer is transported to the faerie world found in the chansons de geste, where modern physics interacts with magic in delightfully strange ways. One might dismiss that novel by saying that it is not science fiction at all but fantasy. How then should one classify his dozens of "puzzle" stories, in which a known scientific law is used as the solution to an obviously contrived problem? Readers of Anderson's "The Three-Cornered Wheel" do not really expect future astronautical entrepreneurs to be stranded on a world where they must transport a large object over bad roads, but find that there is a religious taboo against constructing or even drawing circles and round wheels.

Science fiction does not predict the future. It does, however, often succeed at technological forecasting. Although the success record of SF authors as a group is not startling, the best of them are most likely on a par with such professional forecasting institutions as the Hudson Institute or the U.S. government's Office of Technology Assessment. This should not be surprising because many science fiction writers have training in technological assessment that equals or surpasses that of the professional forecasters and draw on the same source materials. The famous "secret" letter to U.S. President Franklin Roosevelt in which Albert Einstein pointed out the possibilities of an atomic bomb did no more than reflect what was known to anyone familiar with the open technical literature of the time. Einstein's equation ($E = mc^2$) demonstrating the possibility of converting matter to energy had been discussed in physics

journals for more than 20 years, and when the United States began the Manhattan Project other major powers (Germany, Japan, and Great Britain) had already begun research on practical devices for accomplishing what had long been known to be possible in theory.

Sometimes the relationship has been more direct. Given that during World War II such SF writers as Robert Heinlein and L. Sprague de Camp, who were both engineers, had been assigned work on pressure suits that would withstand total vacuum, it is no great surprise that the space suits worn by early astronauts were quite similar to those described in science fiction stories of the late 1940s and early 1950s.

Finally, many science fiction writers routinely maintain close relationships with the technological community. Thus what is often seen as a startlingly successful forecast is in fact no more (and no less!) than the first popularization of an already accomplished laboratory breakthrough.

Discussion of science fiction as a literature of prediction can easily founder in a quagmire of definitions—not only of SF but of prediction itself. As an example, most writers active between 1940 and 1960 were interested in the then-infant computer sciences. Many important stories included very large and very complex computers. The science fiction machines—even those of the rather distant future—generally resembled real computers of the time: enormous things, occupying many square feet of floor space and filled with thousands of vacuum tubes and hundreds of thousands of discrete electrical parts. What was not foreseen, either by science fiction writers or professional forecasters, was that within two decades computers would become not only vastly more complex and capable but also small and cheap. Does one claim a successful prediction of powerful computers, or failure because science fiction has not to this day dealt with the consequences of widespread distribution of computers and information systems?

To quibble over this question is to show the futility of such a discussion. At best, science fiction has no more utility in either predicting the future or accurate technological forecasting than does popular nonfiction science literature. Its claim to significance must rest on something more substantial, as in fact it does. Rather than merely predicting the future, SF can make a good claim to shaping it. In fact, it can claim, with justice, to be among the most influential literature of the century.

# A BETTER PROSELYTIZER

One important influence of science fiction becomes obvious at any major scientific convention or event: the career choices of many, perhaps a majority, of the scientists and engineers present were profoundly influenced by early exposure to science fiction. They may no longer read science fiction—modern science is a harsh mistress and leaves little time for amusement—but thousands of scientists first became fascinated with science and technology through science fiction. Indeed, it would not be hard to make the case that without the stories of Heinlein alone, the already difficult task of aerospace company recruiters would be impossible.

It is generally agreed that the world needs a steady supply of good engineers and scientists. Unlike the liberal arts, hard-science courses are unforgiving. Failing science majors soon find it of no use invoking cultural relativism or explaining that they "have an open mind" and thus do not accept the current theory of the differential calculus. A physicist may advance to a point at which his refutation of Einstein's theory of gravitation will be taken seriously, but not without first having demonstrated to some instructor in his academic past that he understands general relativity. In contrast, the social sciences and liberal arts have significantly less intractable content, and in those fields the ability to argue one's case can be as important as scholarship. The hard sciences need bards to sing their praises if they are to attract new converts, and science fiction serves that need.

It is more difficult to measure the influence of science fiction on the lives of individuals in fields apart from the sciences, although that influence cannot be negligible and may be important. However, the spell that SF casts over a small but significant population segment is both direct and nearly total. The phenomenon of science fiction fandom is unique. No other literary genre has developed such a large and well-organized cult; nor is there another genre in which such routine, massive, and direct contact exists between authors and readers. Moreover, for many SF readers there is a period—sometimes a few years, sometimes decades—during which science fiction is the only literature read.

The few studies attempting to characterize readers of science fiction unanimously show that the majority are considerably above average in both intelligence and potential social influence. While those who long remain total addicts to SF seldom have great influence outside SF fan organizations, the same is not true of those merely temporarily addicted. Many of today's scientific, academic, political, business, and social leaders literally lived in science fiction story worlds during their adolescence, and although the stories of that era seem today almost hopelessly conservative, they were thought radical in their time. To many of those young readers the worlds of SF were more real and more natural than the actual world in which they grew up. It is small wonder that they have been willing to act as midwives in creating those worlds.

## A VOICE IN THE WILDERNESS

A second, more profound influence is easily seen but difficult to pin down: the preparatory impact of SF on both the general public and the scientific community. Science fiction, wrote Future Shock author Alvin Toffler, "widens our repertoire of possible responses to change." It does this by "dealing with possibilities not ordinarily considered." Marshall McLuhan stated in *The Medium is the Massage* that the problem of modern man is "to adjust, not to invent," and that science fiction will "enable us to perceive the potential of new technologies." Science fiction is needed to "find the environments in which it will be possible to live with our new inventions."

One excellent example is the U.S. space program. As late as 1955 few people believed they would live to see humans reach the Moon. Most had grown up in times before widespread use of electricity, and many were convinced that they had "seen the future" with the initiation of scheduled airliner service. But during the 1950s a major boom occurred in science fiction. A dozen magazines sprang up, although most of these flourished only briefly. Hollywood made dozens of SF movies, most banal and some dreadful but all fairly popular. For the first time it was possible to buy whole books of science fiction, for SF outside magazines was almost nonexistent prior to 1950. Television made its contribution with *Captain Video*, *Space Patrol*, and other such series. Eventually the demand for science fiction was so great that a great deal of very low-grade material was published and filmed. Nevertheless, the impact had been made. The idea of manned space flight was no longer "far out," "weird," and "not for our lifetimes." It was in the air, something thought of every week, and not farfetched at all compared with what movies were offering. After all, if scientists could unerringly save mankind from Godzilla, giant ants, and other Hollywood creatures, what could stump them?

Although probably impossible to prove, it can be argued that without the preparatory influence of SF, the U.S. Apollo program would not have been possible. Certainly President John F. Kennedy's announcement of a manned Moon landing before 1970 would not have been well received. He would have been thought a frivolous dreamer, not an imaginative leader. Moreover, much of the popular acceptance of the Apollo program must be laid to science fiction and definitely not to the technological community. When Kennedy made his announcement in 1961, most aerospace engineers were agreed that his goal was mere moonshine. In fact, the more closely they were attached to the space program, the more likely they were to insist vehemently that getting to the Moon in nine years was impossible.

The same kind of preparatory role can be seen operating today. Throughout the history of science fiction, writers have offered hundreds, perhaps thousands, of stories of the first contact between mankind and intelligent life of extraterrestrial origin. In some of these works the setting is the past, present, or future Earth; in others it is in space or on another planet. A few authors have even finished in fiction the real-life drama of scientists who presently listen to radio noise from space for broadcasts from other civilizations among the stars. If on some future day Earth's inhabitants discover they are not alone in the universe, they will be better equipped to deal with the biological, social, philosophical, and religious implications for having experienced them in the literature of science fiction.

Although science fiction's preparatory role is most dramatically illustrated in the field of technology, there has been a more subtle, but perhaps deeper, effect on social relations. It is impossible to measure the actual contribution of such stories as Anthony Boucher's 1943 novelette "Q.U.R." Ostensibly about robots and robot psychology, the story presents a black man as president of a World Federation of nations. Heinlein's novels written for a juvenile audience regularly employed women as scientists and engineers. Although science fiction legitimately can be faulted for retaining accepted stereotypes and for lacking sufficient boldness in asserting sexual and racial equality, as early as 1955 Heinlein presented Captain Helen Walker, a soldier of the Imperial Army, in *Tunnel in the Sky*. The influence of such stories in shattering cultural stereotypes can certainly be exaggerated, but it should not be underestimated.

## EVERYDAY MIRACLES

Closely tied with the preparatory function of science fiction is its tendency to change people's expectations, sometimes directly, sometimes in rather subtle ways. Moreover, this type of change interacts with science and the real world.

As an example, one of the least studied but most important results of the Apollo program was that those involved learned how to manage incomprehensibly complex tasks. Hundreds of thousands of people worked on thousands of separate projects—some of which involved discovering how to do things previously not possible—and the products of all this activity were brought together at a single time and place to produce a result. This degree of coordinated activity was unprecedented in human history. The only activity remotely as complex has been war, and wars have hardly been famous for good management. The invasion of Normandy on D-Day, 1944, may have approached Apollo in numbers involved but hardly compares in technological complexity. Apollo was unique; moreover, it was on time and very nearly within a budget set years before anyone knew how the mission would be accomplished.

Yet, although the Apollo program was a milestone in history, the difficulty of the task has not been appreciated by the general public—nor, indeed, by science fiction, which had usually shown the first flight to the Moon as a fairly simple accomplishment, sometimes as a backyard project. Although science fiction may have made the flight possible, by underplaying the difficulties it also diminished public sensitivity to what had been done.

The result of this interaction between public expectation, often shaped by science fiction, and scientific accomplishment has been fairly consistent. The public expects miracles and cannot understand why they are not routinely forthcoming. Science fiction encourages people to dream, while the knowledge explosion leads them to demand that the dreams come true.

Of course not all such demands can be ascribed to the influence of science fiction. It has long been an insistence of Western civilization that nature adapt to mankind, not mankind to nature. The biblical book of Genesis tells man to subdue the Earth and have dominion over every living creature. Although there is a school of SF that explicitly exhorts human beings to live in amity with nature, the vast majority of the literature disagrees.

It is noteworthy that the optimistic hard science story of the individual triumphant is popular chiefly in the U.S., with its long tradition of "American know-how" and inventive heroes like Benjamin Franklin and Thomas Edison. It is probably also significant that SF's dramatic rise in popularity during the 1930s and 1940s came during a period when mainstream literature concentrated largely on stories of men and women destroyed by an impersonal society. (Interestingly, a lot of science fiction stays in print for a very long time; even an average SF novel is likely to be available years after many Pulitzer prizewinning works have become unobtainable. Again, one is tempted to ask which is mainstream.)

## TURN BACK, O MAN

Another indirect influence of science fiction is as warning. It is an old and honorable tradition: some of the best-known SF is pure jeremiad; for example, George Orwell's *1984* and Aldous Huxley's *Brave New World.* The generic type is what Heinlein once called the "If This Goes On" story: take a current trend, carry it to extremes, and show a society—usually a dystopia, i.e., a perverted and malevolent utopia—built from the results. There are thousands of examples of stories warning against hundreds of trends, some significant, some utterly trivial.

It is impossible to know just how influential such stories have been. For instance, if Western society is not in fact moving toward a world dominated by advertising agencies, how much of its safety has been due to the warning delivered by C. M. Kornbluth and Frederik Pohl in their 1952 classic, *Gravy Planet*? Nor is such a question absurd; at least some of the present skepticism toward the media appears strongly influenced by science fiction. Even more specifically, one may wonder just how much influence Poul Anderson's popular and prophetic 1954 novelette "Sam Hall" had on passage

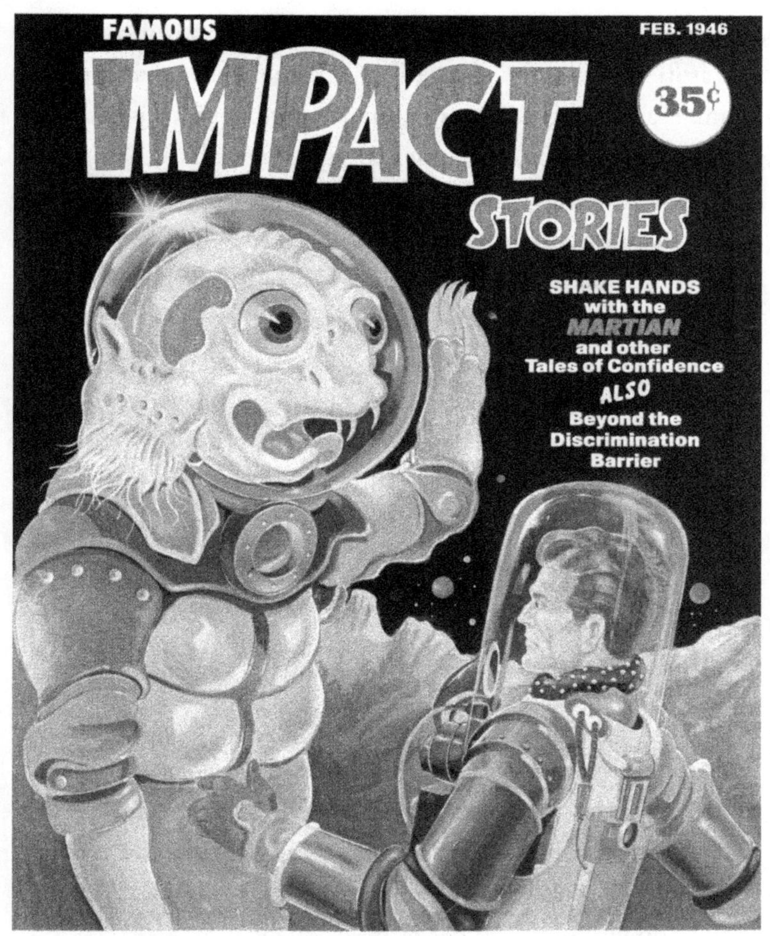


in the U.S. of the Privacy Act of 1974. This legislation regulates the dissemination of information about individuals that has been collected in government dossiers and permits persons to see their own files.

It might be argued that society would be the same today if the above two stories and many other SF jeremiads had never been written. Certainly it is difficult to show the influence of any single story or book, or even of an important author. Yet it is also probable that attitudes have been changed by their cumulative influence. For example, an almost uncountable number of stories showing the grim consequences of war in the nuclear age were published at a time when military and civilian policymakers still believed wholeheartedly in nuclear war as an instrument of national policy.

Although the majority of science fiction has presented science and technology as beneficial, there is a very strong countertrend denouncing science as Faust's bargain. Often those who cry warning have outshouted the larger number of bards who act as technology's harbingers, at least in the opinion of literary critics. Whether the dystopian theme produces better works than the more traditional stories of triumph is debatable, but certainly the gloomier works are more likely to win academic critical acclaim. Indeed, some critics go so far as to say that nothing can be literature that does not recognize man's fallen state.

Although fewer in number, there are also jeremiads warning against the rejection of technology and depicting societies that have sunk hopelessly into misery as a result of foolish attempts to "return to nature." This type of work has not often succeeded in gathering popularity or literary acclaim, probably because protechnology authors do not find such societies interesting, while antitechnology factions do not find them believable.

## WHO SINGS TO THE BARDS?

If science fiction or, more accurately, science fiction writers have a significant influence in shaping the future, what influences them? Perhaps the most consequential factor is fandom. SF fans are important far beyond their numbers or economic impact. It has been estimated that all of fandom—everyone who regularly reads one of the SF amateur publications called fan magazines or "fanzines," plus everyone who attends science fiction conventions—does not number more than 25,000. Whereas this figure would represent a very respectable sale for a hardbound book, it is well known that no large number of fans buys hardbound books; most wait for the paperbacks. Thus, even if an author sells a copy of every paperback he writes to every fan in existence, but to no one else, the author will starve. Simple economics dictates that SF writers must appeal to a larger audience and that, if fans' tastes conflict with those of the general public, on economic grounds the fans ought to be ignored.

Yet few authors do this, for the very good reason that SF fans are usually the only readers an author meets, and it would be an unemotional writer indeed who could ignore the preferences and sentiments of hundreds to thousands of readers. The life of the average professional writer is an unnatural one. He spends many waking hours in a room alone save for beings of his own creation. It is no wonder that many writers seem unfit for social intercourse. And while this situation may apply to any professional novelist, it is even more descriptive of the science fiction writer, who is not often invited to academic parties and in general is not accepted in mainstream literary circles, but who will always be welcomed by fans. The temptation to please organized fandom is extreme.

If fans are important to science fiction writers, so are other science fiction writers. The late author Richard McKenna once described a gathering of SF writers as "a trapper's convention." Only those who have served the lonely hours on the trap lines are full companions in the order. Certainly SF writers form a close-knit community unique among laborers in the literary vineyards; no other genre has anything faintly resembling their fellowship. Most science fiction writers have received substantial assistance from some of the most

famous names in the field, and there is a long-standing tradition that such aid should be "paid forward" to newcomers. New developments in science, even new story ideas, are regularly exchanged. While some SF writers delight in outraging their colleagues, few can entirely ignore them. Most will admit to their influence.

Thus a strange phenomenon has occurred: the creators of the future have in large part created themselves. Although a substantial minority of SF writers have not been fans, they are often drawn into the SF community after they become writers. Fandom and other writers are the major influences on science fiction. Science fiction created fandom. The serpent eats its tail.

## THE LAND OF DREAMS

There have always been bards. They have often held high status. They have sometimes been more important than kings. In ancient days they roamed the land, seeking listeners who would feed them, searching for the campfires of some wandering band, approaching to say, "If you will carve me a slice from that roast and fill my cup with wine, I'll tell you a truly marvelous adventure in a land where men fly, light burns cold like the firefly, and wagons move without beasts to draw them."

Then as now they hoped to sing for more than supper; they imagined that their songs might sway kingdoms and powers and armies. Sometimes they did. Yet, was that their true work? Or is the true work of the bard to dream and, having dreamt, to tell his dreams? Then as now they could not know their influence. They could only tell stories. But nothing happens unless first a dream.

# 20/20 VISION

In 1971, Jerry Pournelle began assembling a collection of short stories by notable writers of science fiction. It was intended, as he said in his Preface, to be a book of predictions:

> *It is not, of course, primarily a test of the power of science fiction writers to predict the future. It is, after all, intended to be entertainment. At the same time,* 20/20 Vision *gives us a unique opportunity....*
>
> *The ground rules on this book were simple. Each author had to write a story which he truly believed could take place in the real world during the year 2020. There were to be no benevolent Alpha Centurian social scientists landing on earth to solve all our problems—unless, of course, the author really thinks the Alpha Centurians are coming in the next fifty years and that they'll give a damn about saving us when they get here.*

The book, eventually published in 1974, contained stories by Poul Anderson, Harlan Ellison, Larry Niven, A. E. Van Vogt, Norman Spinrad, Ben Bova, and a few others. Dr. Pournelle led with the Introduction which follows.

# DO WE LIVE IN A GOLDEN AGE?

When I set out to gather, commission, and bullyrag authors for the stories in this book, I had no idea what the theme would turn out to be.

Surprisingly, as soon as I had collected them, one connecting link fell into place immediately: Do we now, or will we soon, live in a Golden Age?

The Golden Age theme has been popular throughout all recorded literature, but it got a special boost in the writings of the early Greek philosophers. From there it inevitably wormed its way into the literary mainstream of Western Civilization.

It's interesting to speculate on where the idea of a Golden Age came from. I have recently been doing a lot of research, and I think I know where the Golden Age idea originated, at least among the Greeks and peoples of the Mediterranean. It started with Atlantis, and that all by itself should gladden the heart of every good science fiction reader.

There's been a lot of literature on Atlantis. At one time when reporters were asked what the five biggest stories would be that they ever could imagine, they rated the reemergence of Atlantis at the top, even above the Second Coming of Christ. That was forty years ago. Interest in Atlantis then dwindled, until recently; and now that, as I believe, Atlantis has been found, it wasn't so big a story after all. Still, it does tell us something about Golden Ages.

According to archeologists Atlantis may have been what the people we know as Minoans called their empire. We have no direct evidence of this, but we only adopted the term "Minoan" because we had to call the Cretan Thalassocracy something. Using "Minoan" is a bit like saying "Edwardian" for the civilization in England from 900 a.d. to 1315 a.d., a period from the middle of the Anglo-Saxon invasions to Runnymede. Still, there was a king named Minos, and the name was probably a title like "Pharoah"; and for lack of something better, why not "Minoan"? Only now, I think, we've got something better.

Atlantis means the land or people of Atlas, who was mythologically a son of Poseidon—and we know that Poseidon the Earth Shaker, Poseidon the lover of horses and most especially bulls, Poseidon, god of the sea, was held in special veneration by the Cretan Thalassocracy. Besides, the similarity between the Bronze Age civilization Plato described as "Atlantis" reminds us on every line of what we know of Bronze Age Crete.

The Golden Age of Atlantis, then, was a time when the Cretan navy patrolled the seas, rescuing distressed sailors and suppressing piracy; when people lived in cities without walls and apparently could live their lifetimes without participating in war.

It was a time of plenty, with bronze tools, gold jewelry, lovely pottery, free trade on the inland sea, frescoes in private houses, and a code of laws displayed to the people by Talos, the Bronze Man. My own speculation, by the way, is that Talos was the title of the commander of a bronze-armored elite guard who escorted what amounted to a court of assizes on tour throughout the realm of Crete. If you like something better, you're welcome to think up your own.

Somewhere between 1500 B.C. and 1400 B.C., the Cretan Empire was partially destroyed by a sudden calamity. There is plenty of evidence to show that this came at about the same time as the first eruption of Santorin, a still-active volcano some sixty miles from Rnossus, the capital of Crete. While the Cretans were digging out from under the ashes they were conquered by Mycaenean Greeks who might well have been Athenians under a leader named Theseus. There's no direct evidence for this, but why not?

Then, some thirty to fifty years later, Santorin erupted again, this time with a force estimated at six times that of Krakatoa; and the Empire of the Sons of Atlas was washed over by tsunamis which carried volcanic pumice stone to levels a hundred feet above the sea. The Thalassocracy was ended forever.

According to Plato's account of Atlantis, the Athenians had already won their war against the Atlantean Empire, and when Atlantis sank into the sea the Athenian army was lost as well....

What this has to do with the present volume is that we can, if Atlantis really was the Golden Age, see just what later men thought was the epitome of civilization. We can dig it up, read its records, look at its everyday life, drink in its accomplishments, and mourn its passing.

And we can wonder if future archeologists will treat us as kindly. According to a number of very serious studies we now live in a Golden Age, one that won't be repeated for a very long time—if ever.

Just how we'll lose what we have is a matter for argument.

Conservatives think that "moral degeneracy," an increasing tendency toward statism, and corruption of the ancient virtues of the American Republic will bring us down to the barbarians of communism (or the left) just as similar factors betrayed Rome first to the Princes, then to the Dominate, and finally to the Goths, Vandals, and Lombards.

Others, not of a conservative persuasion, think we're headed for a fascist state, repressive for the sake of repression, dedicated to the proposition that anything preserving what we've got is all right no matter how brutal we must be to do it.

In fact, about the only thing most commentators of all political persuasions can agree on is that something drastic is going to happen.

Back in the thirties Albert Jay Nock, who believed we were passing into a New Dark Age, was so little noticed that he entitled his masterwork The Memoirs of a Superfluous Man. In those times nearly everyone believed that history was the record of continuous and inevitable progress. Today, Nock's thesis is taken a bit more seriously.

Dr. Isaac Asimov has a whole list of reasons why we're all doomed, but they can be summed up in his famous vision of the future: Crowded! This is a view shared by many professional futurologists. A recent issue of The Futurist states flatly: "Man appears to be heading toward a calamitous Day of Reckoning. Unless his rapidly growing population and expanding industrial capacity is somehow brought under control, the earth's natural resources will be exhausted and the environment so polluted that the world will no longer be livable."

Nearly every professional study of the future predicts a world population in 2020 greater than twice that at present; some go as high as four times that—fourteen billion people, all crowded onto this "spaceship earth."

Needless to say, the same studies then show a rapid decline in population—a biological die-off. They also show food per capita increasing slowly until about the year 2000, then falling rapidly as population gets ahead of pest-proof storage bins, miracle rice, fertilizer, sea farms, nuclear powered desalinization of water, and all the other recent technological marvels that together make up the "green revolution."

Pollution projections start with the present base level shown as almost nil. Pollution rises steadily beginning around 1980 until it, too, peaks, and the die-off begins in the middle of the next century. Meanwhile, the supply of natural resources has been falling since 1900 and will continue to fall at an accelerating rate.

Put it all together onto a curve. Call that curve "Quality of Life," or "Material Standard of Living." Whatever you call it, according to some of our best professional projections it has already peaked and can go nowhere but down. We live, in this estimate, in a Golden Age which will be remembered fondly and nostalgically if at all.

No matter where we look, then —to politicians of very different ideological persuasion; to futurologists; to ecologists; even to science fiction authors —we see little to hope for. Perhaps, indeed, we live in the Golden Age.

Just how inevitable is the grim future some forecast for us? Can nothing save us?

Well, for one thing, that depends on what we mean by "save"; and perhaps even more it depends on what we mean by "us". If we mean the people of the United States, salvation may be at hand. Provided that we avoid nuclear war, the U.S. population is presently under control. Best estimates now indicate that it will continue slowly to rise for a while, will peak somewhere around the year 2,000, and then gradually fade back, reaching in 2020 very nearly the present level where it will thereafter remain constant.

Meanwhile, the average age of the U.S. population will creep up and up, leaving proportionately fewer people in the work force, so that if we don't get cracking with new technology to increase the productivity per worker there's going to be a good bit less for all of us when we retire. Our present Social Security system is, of course, based on the assumption of continuously rising productivity; it taxes the work force to support those retired from it, and at present levels of taxation and production those who are right now forty years old and under cannot be accommodated when they retire.

Still, there's no reason to think we can't increase the productivity per worker; in fact, there are fears in the opposite direction, that automation will so increase output per employed person that we'll face increasing unemployment of the potentially productive. Which, then, is the greater fear: unemployment or insufficient work force?

Without giving detailed reasons, I think that for the United States, Japan, Western Europe, and the U.S.S.R. technology will be able to keep pace with population trends including the aging of the work force; but when we turn to the rest of the world which will, after all, contain far more than half the people in the year 2020, the picture is much grimmer.

At present rates of energy and natural resource consumption there is not a ghost of a chance that the capital investment needed to bring the "developing nations" into a high-energy/high-technology society—not like ours, but, say, like the one we had in 1940—will be forthcoming. If you take capital investment per person in the U.S. and multiply that by the number of people worldwide in 2020, you come up with a number so big that it isn't worth reporting. It can't happen, and that's that.

(What is the number? Well, present estimates show that it takes about $30,000 capital invested to create one job in the U.S. $3.0 x 10 x 10 people = $3 x 10 or three quadrillion dollars; a substantial sum. For 1940 investment levels it's still up near a quadrillion.)

Moreover, there just isn't enough iron ore or oil or coal or any other traditional natural resource to allocate enough per capita worldwide to construct a high-technology civilization like ours, or Western Europe's or Japan's; it's going to take quite a lot even to do it for the U.S.S.R.

So where does that leave us? There are several ways to go, and that's the problem with "scientific" predictions of the future.

Whenever scientists predict, they have to recognize that their own actions will influence the outcome of the game; and even if nobody takes their predictions seriously, still, politicians and other decision makers have a habit of acting quite unpredictably.

But here are some of the ways we could go. First, the Western democracies could voluntarily decrease their standard of living to provide resources—massive amounts of resources—for the rest of the world. Discounting the vast possibilities for mistakes and maladministration of these gifts—and the history of our foreign aid program does not entitle us to discount the possibility at all—

there aren't too many politicians who want to run for office on that platform.

There could be a revolution in the Western democracies which their critics say aren't very democratic at all to begin with. All well and good, although the objective probability of that is pretty low; but granting that it happened, the record of generosity of revolutionary societies is really rather poor.

Even assuming that the revolution left the productive mechanism intact so that massive worldwide technological aid and investment was possible, a restructured U.S. society "truly responsive to the will of the people" would probably find the people's willingness to deprive themselves a bit lower than ideal.

Well, the technologically backward areas of the world might force the high-energy civilizations to divvy up. Since the ways of applying coercive force are directly dependent on the technology available to the contending sides, this doesn't look very inviting; the chances of India invading the U.S., Britain, or the U.S.S.R. are slim.

China might have greater success at blackmailing the Japanese or the Soviets, but when you really get down to brass tacks their chances will depend on the willingness of those two giants to defend themselves; and after the experiences in Hungary and Czechoslovakia I myself shouldn't want to bet on the soft-heartedness of the Soviets. Japan alone can't do that much for China anyway, although she could do a lot.

Where else might we go? Even with new birth control technologies—let's postulate one thoroughly acceptable to all the churches which now oppose present techniques—the populations of the "developing" nations are going up and up.

Nothing we could do would lower the birth rates worldwide without a considerable time lag; the "green revolution" provides enough food for subsistence survival of an enormous increase in population; and medical technology is easily exportable and in fact is being exported even as I write this.

Increased longevity contributes more to short-run population expansion than birth rate, anyway. The real spurt in Western populations came as a result of the discovery by a Hungarian physician in the last century that if doctors and midwives washed their hands before delivering babies the mothers didn't die of "puerperal fever" and thus survived to have more children. The doctor was locked up by his fellow physicians as a madman, by the way; they simply wouldn't believe that they were the cause of the mysterious childbed fevers...

I see, then, only one way out of the dilemma. We've got to develop a whole new kind of technology one that doesn't depend on natural resources at all and doesn't take the kind of capital investment we used to build our own civilization.

Such a technology is possible. The Oak Ridge National Laboratories, for example, have developed a whole program for "instant industrialization" of such barren areas as the Rann of Kutch, the Namib Desert, Sinai, and the Egyptian Red Sea coast.

These make use of fast breeder fission reactors to produce electric power which can be used for desalinization of water. The water supports crops, and the bitterns left from the sea water are "mined" for chemicals and minerals. The waste heat or "thermal pollution" of the reactor is useful energy for the "mining" process; while the breeder construction continuously makes plutonium fuel from now-useless U-238.

There are other possibilities, of course, but they all depend on finding sources of energy other than the traditional fossil fuels. The evidence is overwhelming that a chemical-energy civilization simply cannot be made worldwide, nor sustained if built. We have one energy source—fission—now. Will we find something else, and will we find it quickly?

Given fission technology already on the shelf we could save the Golden Age. The investment requirements are large, but not on the order of quadrillions of dollars; hundreds of billions won't even be required if the wastage factor is kept to reasonable levels—and provided that "concerned" amateurs don't bring technology utilization to a screeching halt.

Sure, it would probably be "better" to wait for fusion technology before embarking on worldwide development schemes—but do we have time? It took thirty years to go from Fermi's pile in the University of Chicago squash court to useful electric power from fission, and there is little reason to believe that fusion or solar screens or any of the other exotic ideas kicked around by science fiction writers and ecologists will do any better. For that matter, not one of those technologies is on the squash court yet. And less than thirty years from now is the year 2000.

That other doom, the population explosion, doesn't worry me as much as it seems to concern some of our doomsayers. I don't think, for example, that we shall all go mad because of the number of people per square mile that we can reasonably expect in 2020. The population per unit area of the Netherlands, as an example, is much higher than we'll ever live to see even throughout Europe; and as anyone who has visited Holland knows, it's quite a sane and charming place.

Furthermore, the record of industrial society is clear: The higher the technology and the more secure the population, the lower the population growth. That could be because it's easier to persuade people of the danger. It could even be that only high-energy civilizations look beyond the next few years and give a damn.

I think, then, that the year 2020 could look a lot like the year 1970, only more so and all over, with some of those nice touches that science fiction writers used to add back when we were a more optimistic breed. Antipodes rockets, great wealth per capita, personal computer terminals, pollution-free rapid transportation nearly everywhere at low prices, and all the rest are certainly within our technological capabilities.

Of course, "more of the same" scares some people silly, too, and horrifies others. That's one area imaginative science fiction writers can help with: It's easier for an SF author to have a vision of a more pleasant future organization of society than it is for the more traditional scientific prognosticators, who by and large are stuck with trend analysis and projection. On the other hand, we can't ever forget the rather grim technological realities: Any kind of "pastoral society" based on early twentieth century technology condemns a very large number of human beings to a very unpleasant death.

On the other hand, wealth for the U.S., Europe, Japan, and the U.S.S.R. is within our capabilities no matter what we do about the rest of the world. Unless we act fast that might be the best we can hope for. But whether we can keep that wealth while surrounded by a vast sea of people facing famine and a biological die-off is something else again; and what we may have to do to ourselves to want that kind of life is perhaps even more frightening.

It could be that we'll be richer in 2020 than we are now—and still see 1970 as The Golden Age.

# APPENDIX: BIOGRAPHY OF RON VILLANI

This biography of the artist who illustrated the article "Bards of the Sciences" was kindly provided by Vanessa Villani, his daughter.

---

Ron Villani was born in Chicago, Illinois in 1939. He had a prodigious imagination and wild creativity that flowed out of him uninterrupted his entire life. His ideas found their way into stories and onto canvases, sketchbooks, napkins and any fragment he could get his hands on. He possessed a playfulness and sense of wonder that never left him. A quiet observer of the world around him; his artwork expressed the beauty he saw there, as well as a terrifying reality. He was a perennial student of information and possessed a quiet humility that caused him to be less interested in his own ideas and more in people or subject matter that could lend something to his own life and art. His interests included classical music and opera, natural history and science, the circus, science fiction, rockets and robots,Greek and Roman mythology, medieval lore, motorcycles and machines, World War II planes and imagery, and most of all, his family.

He exhibited a precocious artistic talent as early as elementary school and took his first drawing classes at the School of the Art Institute of Chicago at the age of ten. After graduating from Fenger High School he received a gymnastics scholarship to the University of Illinois at Champaign Urbana where he stayed several semesters before riding his motorcycle to California to seek adventure and take writing classes at UCLA. Eventually, he returned to Chicago to attend the School of the Art Institute working as a life model to pay his tuition. Upon earning his Bachelor of Fine Arts degree in 1963 he received the prestigious Ryerson Foreign Traveling Fellowship as well as a scholarship to the Skowhegan School of Art in Maine which he attended that summer.

Ron then advanced his draft in the United States Army. He was appointed to the position of Graphic Design Director in the Army's Publications Department at Fort Belvoir outside of Washington, D.C. The work he produced there made him indispensable; so much so, that his commanding officer called off his planned tour of duty in Vietnam. After serving for two years he proceeded to make use of his Ryerson Fellowship and travel to France, Italy, Spain and other countries across Europe. He was happily accompanied by his wife, Judy, a fellow graduate of the Art Institute of Chicago.

His professional career as an illustrator, designer and art director spanned six decades. He was the design supervisor at Encyclopaedia Britannica Inc. He also did freelance work, illustrating for advertising agencies, design firms, record companies, national publications, and museums. Ron served as an art director and produced artwork for Playboy, Harley Davidson, Apple, Audi of America, Crate & Barrel, Target, Sports Illustrated, Anheuser-Busch, Navistar International, McDonald's, The Museum of Science and Industry, The Field Museum, The Kenosha Public Museum, Unical of California, Mercedes Benz, Jaguar, Chicago Teachers Union, Klutz Press, University of Chicago Press and Scott Foresman just to name a few. His artistic styles were so nimble and varied that eventually, in addition to Ron Villani, he went under two different pseudonyms, "Buc Rogers" and RV2 to span the scope of his work. Upon semi-retirement he was able to devote himself to his passion of painting. His artwork was represented and received awards at various galleries and museums throughout his lifetime in Chicago, as well as in prominent collections on the East Coast and throughout the Midwest.

As an artist, Ron was focused on creating works of fine art that reflected his awareness of social issues, his keen sense of humor, his diverse interests, and his technical prowess in virtually any medium. His take on life was entirely unique, and his art was fueled by a fascination with science fiction, the unaffected, the absurd, the dismissed and the epic. Drawing was his fundamental discipline, and he was an extraordinarily skilled draftsman who could draw anything in whatever style was required. His skills in realistic representation were impressive, to say the least, but he easily shifted from realism to more imagination-based representational modes. His paintings communicate not only through their bold subjects but also through his handling of popular culture references to create a world that is seductive, familiar, playful and complex. Visceral figural distortions, graphic stylizations, virtuoso draftsmanship and powerful social commentary tinged with humorous and macabre elements inform his artwork. Ron was best able to represent his biggest hope for mankind through the lens and expressions of humanoid robots which were a constant theme in his paintings and illustrations. The last line of his artist's statement proclaims "Maybe robots are the answer."

His unique vision and prolific creativity was a constant throughout his lifetime. Ron had endless plans for drawings, paintings, and projects in the works that he most certainly would have brought to fruition had he not died unexpectedly on November 13th, 2017.

CPSIA information can be obtained
at www.ICGtesting.com
Printed in the USA
BVHW081016270321
603566BV00007B/1123